ELECTRICAL ENGINEERING DEVELOPMENTS

SMART MICROGRIDS

NEW ADVANCES, CHALLENGES AND OPPORTUNITIES IN THE ACTUAL POWER SYSTEMS

Electrical Engineering Developments

Additional books in this series can be found on Nova's website
under the Series tab.

Additional e-books in this series can be found on Nova's website
under the e-books tab.

ELECTRICAL ENGINEERING DEVELOPMENTS

SMART MICROGRIDS

NEW ADVANCES, CHALLENGES AND OPPORTUNITIES IN THE ACTUAL POWER SYSTEMS

HUSSEIN M. KHODR, Ph.D.
EDITOR

nova publishers
New York

For permission to use material from this book please contact us:
Telephone 631-231-7269; Fax 631-231-8175
Web Site: http://www.novapublishers.com

Library of Congress Cataloging-in-Publication Data

Smart microgrids : new advances, challenges, and opportunities in the actual power systems / editor, Hussein M. Khodr (Qassim University, Engineering College, Department of Electrical Engineering, Buraidah, KSA).
pages cm
Includes bibliographical references and index.
ISBN 978-1-62808-975-2 (hardcover)
1. Smart power grids. I. Khodr, Hussein M., editor of compilation.
TK3105.S567 2013
621.31--dc23

2013032591

Published by Nova Science Publishers, Inc. † New York

Contents

PREFACE

In recent years electrical power and energy systems have been undergoing unprecedented evolution which has never been observed before. The liberalization of electricity markets, the integration of electrical vehicles, the information technology innovations, the advanced communication systems, the smart-metering and others, are all contributing to major changes in the basic operation of electrical power systems. Consequently, all these aspects have to move the electrical power and energy systems to be more intelligent, providing further benefits, not only for the utilities and power systems operators, but also for the consumers and the society as a whole. This book aims to share, to address and to disseminate in a straightforward way the state-of-the-art research, development and implementation of microgrids to bring together researchers and engineers worldwide from both academia and industry, with the aim of fostering new advances, identifying challenges and opportunities in system operation and maintenance, planning and design among engineers, researchers and stakeholders in this important field of knowledge which is the smart grids.

This textbook consists of six chapters that have been written by relevant academics, researchers and engineers. Most of the chapters include state-of-the-art of algorithms, methods, know-how and examples of applications in academic or industrial problems and their solutions through the most adequate optimization technique. Through the techniques outlined in this book, researchers, engineers and students are encouraged to develop their own simple computer programs to solve the problems they encounter.

Chapter 1 - compares a statistical approach namely Auto regressive Integrated Moving Average (ARIMA) with two learning-based approaches namely Support Vector Machine (SVM) and Extreme Learning Machine

(ELM) on wind speed time series forecasting. The statistics of wind speed time series are analyzed. The details of ARIMA, SVM and ELM are elaborated. An actual off-shore wind speed data is selected for modeling and forecasting. The results show the comparisons of ARIMA against SVM and ELM as well as the comparisons between SVM and ELM on three forecasting time-horizons according to three different error measurements. The need for all these mentioned techniques is because wind power is considered as an important alternative resource for electricity generation. It could be operated in an isolated or connected system. Wind power integration into the power grid is challenging, due to the intermittent nature of wind and its forecasting difficulties.

Chapter 2 - studies the short-term scheduling problem of an isolated microgrid with combined heat and power (CHP) units and renewable energy resources such as wind and solar energies. The considered microgrid can be an isolated power system, an urban area or a large company with heat and power demand. The objective of the short-term scheduling is determining the optimal schedule of different power and heat generation units for supplying the power and heat demands with minimum cost. The technical constraints of CHP generators are considered in the proposed optimization framework. In this chapter, the dual dependency between generated heat and power of CHP units is considered and mathematical models of different CHP types are derived. The effect of uncertainty of wind and solar energy productions are considered using the implemented reserve requirement approach. In this approach a portion of forecasted load and renewable generation is considered as the desirable system reserve. A battery storage is incorporated in the studied system model. Using energy storage systems increases the utilization of the renewable energy resources and the total system efficiency. The proposed model is formulated as a mixed integer nonlinear programming problem and solved using GAMS. An illustrative example is presented in order to demonstrate the applicability of the proposed method.

Chapter 3 - presents an overview on Demand Response (DR) and its application in the context of the smartgrids. Demand Response attains a crucial role in the context of the operations and economy of the smart grids. DR is a collective term for any action taken from the demand side in a proactive or reactive way to cater to the needs of the grid. Although specifically, voluntary load management, load control, peak shaving and peak shifting have been the flag bearing applications of DR, the scope of research for other DR programs and related business models is ample. Smart grids transformed the existing DR programs to fully automated schemes with the inception of open data standards. This chapter provides an overview of DR in all of its aspects covering the concept, the enabling technology, the mathematics involved, the global standards for implementation, challenges and the future research. Interested readers can follow-up for further details by the references provided at the end of the chapter.

Chapter 4 - provides some insights about the motor's control in the new context of smartgrids. Accuratecontrol of dc motors can only be achieved by a well-designed controller such as a closed-loop PID or similar control techniques. The PID controller is mainly employed to control dc motors driving industrial machinery but, its performance highly depends on the gain coefficients. Although the Z-N approach is one of the well-known deterministic techniques to find these coefficients, it is time-intensive. Alternatively, the GA and similar methods are proposed by a limited number of researchers to design the PID controller by identifying gain coefficients according to the pre-defined conditions due to complexity of the Z-N application. In this study, the GA method was used to build the PID controller for the speed control of a separately-excited dc motor. The results were compared with those obtained from implementing the widely-used controller design techniques of Z-N and the PSO for different types of motors. The results have shown that the GA considerably improves the performance parameters of steady-state error, the rising time, the settling time and overshoot minimization with respect to other methods for a dc motor under consideration and it can easily be applied to similar dc motors with the same purpose.

Chapter 5 - proposes some of the new trends to be implemented in smartgrids at medium and low voltage. Requirements to address the specific challenges for protection and automation are identified. These are due to the structure of the electric power grid which is set to change dramatically over the coming years with the implementation of distributed generation and new technologies to improve efficiency and move towards smartgrids. In this sense, advanced distribution automation and protection are introduced along with their network architecture and arrangement of different types of IEDs proposed for smartgrids. These schemes are intended to allow a reliable operation of distribution networks not only under grid-connected condition but also during autonomous operation.

Chapter 6 - presents an optimization model for power distribution. The objectives are to minimize power losses and balance loads among feeders and are subject to constraints such as capacity limit of branches, minimum and maximum power limits of substations or distributed generators, minimum deviation of bus voltages and radial optimal operation of networks. The problem is formulated as a Mixed-Integer Linear programming problem and solved using LINGO. This model can be implemented into a SCADA platform as an online application to solve the overloaded conditions of the networks or to isolate faults, making the network more intelligent acting as a smart network controlled by the SCADA and obeying all the technical constraints. The methodology has been applied to a large real power distribution system with significant savings by reducing active power losses, demonstrating the effectiveness of the proposed methodology integrated with the SCADA system.

Editor
Hussein M. Khodr

In: Smart Microgrids
Editor: Hussein M. Khodr
ISBN:978-62808-975-2

Chapter 1

WIND SPEED FORECASTING: A COMPARISON BETWEEN STATISTICAL APPROACH AND LEARNING-BASED APPROACH

***Ren Ye*[1]*, *P. N. Suganthan*[1†] *and Narasimalu Srikanth*[2‡]
[1]Nanyang Technological University
School of Electrical and Electronic Engineering, Singapore
[2]Energy Research Institute @ NTU (ERI@N)
CleanTech One, Singapore

Abstract

This chapter compares a statistical approach namely Auto Regressive Integrated Moving Average (ARIMA) with two learning-based approaches namely Support Vector Machine (SVM) and Extreme Learning Machine (ELM) on wind speed time series forecasting. The statistics of wind speed time series are analyzed. The details of ARIMA, SVM and ELM are elaborated. An actual off-shore wind speed data is selected for modelling and forecasting. The results show the comparisons of ARIMA

*Email: re0003ye@ntu.edu.sg
†Email: epnsugan@ntu.edu.sg
‡Email: nsrikanth@ntu.edu.sg

against SVM and ELM as well as the comparisons between SVM and ELM on three forecasting time-horizons according to three different error measurements.

Keywords: wind speed, wind power, forecasting, ARIMA, SVM, ELM
PACS 88.50.Mp, 02.50.-r, 88.05.Lg

1. Introduction

Wind Power is a fast-growing energy source that is clean and sustainable. It is an important alternative resource for electricity generation [3]. However, due to the intermittent nature of wind, it is challenging to integrate wind power into the power grid efficiently. Wind power forecasting is required to integrate the wind power produced by wind turbines into the power grid. High precision wind power forecasting is necessary when the penetration level of wind power is high [15].

Accurate wind power forecasting is beneficial for the electrical system. It is an essential process for economic dispatch planning and unit commitment decision if the electrical system consists of wind energy. In addition, it is also used to optimize maintenance scheduling and regulation actions in order to reduce the operation and maintenance costs [11].

Theoretically, wind power and wind speed are related according to:

$$P = \frac{1}{2}\rho A C_p(\lambda, \beta) v^3 \tag{1}$$

where ρ is the air density, A is the area of the turbine when rotating, $C_p(\lambda, \beta)$ is the efficiency and v is the up-wind speed.

Wind speed forecasting is an indirect way to achieve wind power forecasting as shown in equation (1). The additional function approximation will introduce some inaccuracy to wind power forecasting but the literature are still in favor of wind speed forecasting because (i) the wind speed time series data is easier to obtain and (ii) wind speed forecasting can be used for wind farm planning in advance of wind turbine construction [8, 11].

Wind speed forecasting can be categorized based on the forecasting time horizon. Time horizon which is between a few minutes and half an hour is defined as very-short-term forecasting, and half an hour up to 6 hours ahead is called short-term forecasting. Mid-term forecasting is bounded by a time horizon from 6 hours to 24 hours ahead and long-term forecasting is beyond 24 hours [3, 11, 15].

Wind speed forecasting is well studied in the existing literature. There are mainly three approaches for wind speed forecasting: physical approach, statistical approach and learning-based approach [3, 15].

Physical approach uses the meteorological data obtained by physical sensors or satellites to forecast the wind speed of a particular region. A NWP model is widely applied in the real wind power industry. The accuracy is guaranteed by the advanced hardware but the trade-off is the high hardware cost and high computational cost.

Statistical approach employs the historical data to predict the future data. Additional information such as temperature, pressure and wind direction, etc., can also be included as the statistical model's inputs. Some well developed statistical methods are AR, MA, ARMA and ARIMA, etc. The advantage of statistical approach is that it requires less computational power and less resources. However, the disadvantage is that it is linear thus not accurate especially for long-term forecast. The statistical approach is usually applied for very-short-term or short-term forecast.

Learning-based approach is advantageous over statistical approach because of their non-linear natural characteristics. To design a computational intelligence learning-based predictor, there are two phases: (i) supervised learning, and (ii) evaluation on unknown dataset. The most widely used approaches are ANN, decision trees, SVM, and ELM, etc.

This chapter focuses on the statistical approach and the learning-based approach for wind speed forecasting. The objective of this chapter is to compare the performances of ARIMA, SVM and ELM forecasting on the wind speed time series data over different time horizons.

The remaining of this chapter is organized as follows: Section 2. analyzes the wind speed statistics. Section 3. and Section 4. elaborate the details of ARIMA, SVM and ELM. Section 5. discusses the results of wind speed time series forecasting and finally Section 5. concludes the chapter.

2. Wind Speed Time Series Statistics

The statistics of a wind speed time series include the probability distribution function estimation, the auto-correlation function and stationarity. This section will use NDBC station 41001 data obtained in year 2011 on 10-minute average [2] for discussion. The time series is plotted in Figure 1. Notice that there was no data record from mid October to early November thus a straight line is plotted during that time frame.

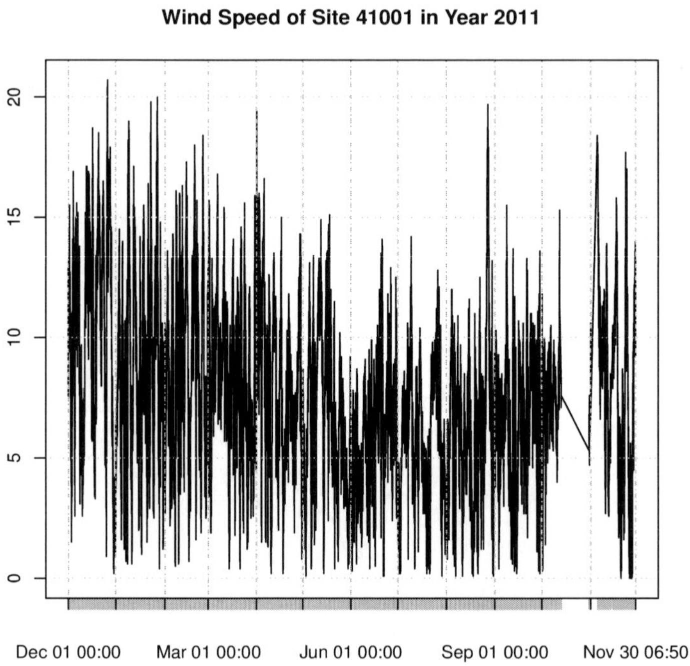

Figure 1. Time Series Plot of Station 41001 Wind Speed in Year 2011.

2.1. Distribution

Empirical results show that wind speed follows a Weibull probability distribution function. A Weibull distribution has a PDF as [14]:

$$f(x;\lambda,k) = \begin{cases} \frac{k}{\lambda}\left(\frac{x}{\lambda}\right)^{k-1} e^{-(x/\lambda)^k} & x \geq 0, \\ 0 & x < 0 \end{cases} \quad (2)$$

where k is the shape and λ is the scale.

In order to fit the actual wind speed time series data to the Weibull PDF, it is firstly to plot a histogram of the time series and secondly to fit the histogram with the maximum likelihood estimation of k and λ. The histogram (stair step, black color) is plotted in Figure 2 and the fitted Weibull PDF (solid smooth, red color) is superimposed on the histogram. The estimated k and λ are 9.5396 and 2.5439 respectively.

2.2. Auto-Correlation

To test whether a time domain signal has correlations between values at different time points (e.g., whether the wind speed at $t = 0$ min affects the wind speed at $t = 10$ min), ACF is introduced. ACF is the ratio of the auto covariance function and the two self-covariances [14]. The auto covariance function is defined as:

$$\gamma(s,t) = \mathrm{E}[(X_s - \mu_s)(X_t - \mu_t)] \quad (3)$$

Therefore, the ACF is:

$$\rho(s,t) = \frac{\gamma(s,t)}{\sqrt{\gamma(s,s)\gamma(t,t)}} = \frac{\gamma(s,t)}{|\sigma_s\sigma_t|} \quad (4)$$

where μ is the sample mean and σ is the sample standard deviation.

However, ACF shows the correlation between data at time s and t with all the data between time points s and t considered, i.e. ACF has a propagation effect from time s to time t. For some applications, this propagation effect will not be relevant but the sole correlation at time s and t is the only focus. PACF shows the correlation between s and t without considering in-between time points [14].

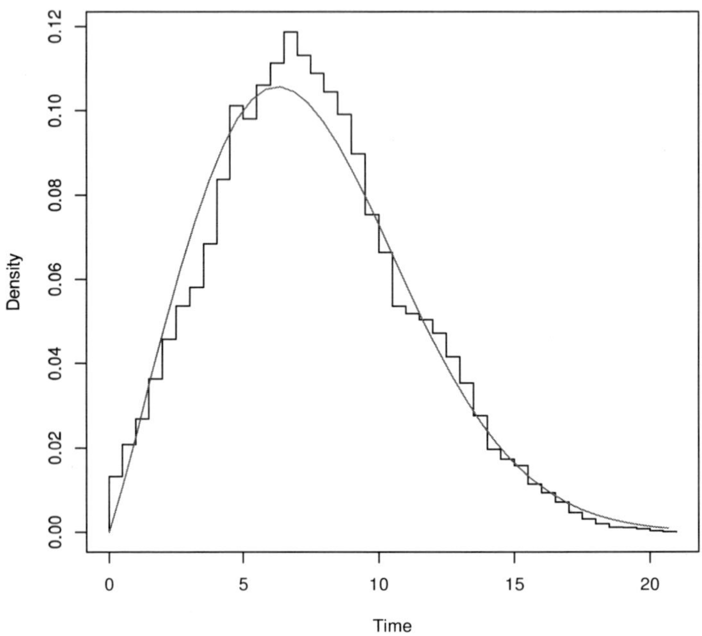

Figure 2. Histogram and PDF Fit of 2011 Station 41001 Wind Speed.

The ACF and PACF plots of the yearly wind speed data is shown in Figure 3. As shown, the ACF decays linearly with respect to lag $h = |s-t|$ and the PACF decays rapidly and become insignificant after fifth lag. The characteristics of ACF and PACF will be discussed in the following sections.

2.3. Stationarity

A time series is said to be stationary if the probabilistic behavior of every values in the original time series is equal to the time shifted version [8, 14]. However, this requirement is difficult to meet, so a weakly stationary time series is mostly used in time series modeling and regression.

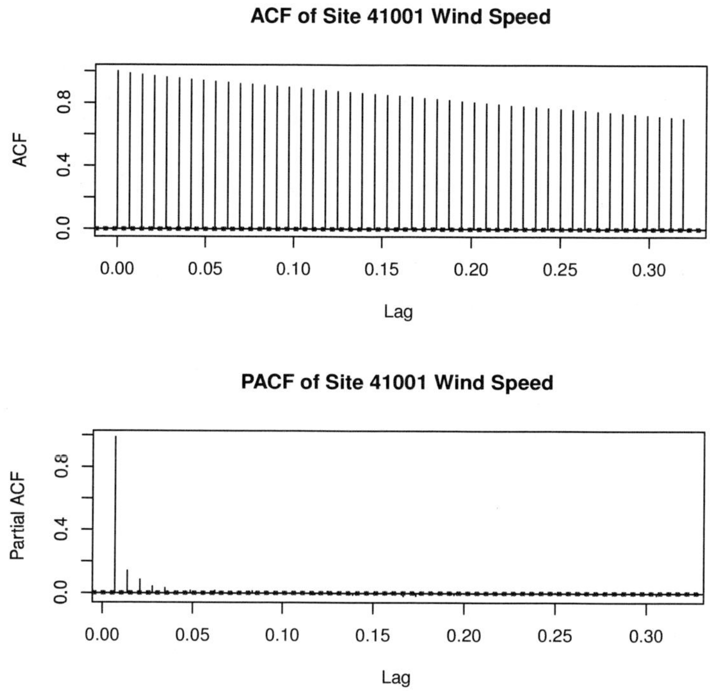

Figure 3. ACF and PACF of 2011 Station 41001 Wind Speed.

A weakly stationary time series has two properties: (i) the mean of the time series is constant and does not depend on time, and (ii) the ACF depends on s and t only through their difference $h = |s - t|$.

A simple way to tell whether a time series is stationary is to check whether the ACF decays rapidly. As shown in Figure 3, the ACF of Site 41001 wind speed data did not decay rapidly, and therefore, this time series is not stationary.

3. Statistical Approach

Time series regression is to fit a deterministic function to the time series data. Traditional time series regression is linear regression, which is insufficient to adapt to the dynamics of a time series. ARMA model is a matured tool for time series regression. ARIMA is an extended version of ARMA which includes a non-stationary-to-stationary time series transformation. This section will progressively introduce ARIMA by firstly introducing AR and then ARMA and finally ARIMA.

3.1. Persistent Model

Persistent model is the simplest model among the models in statistical approach. Persistent model assumes the future wind speed is the same as the current speed as shown in (5). Persistent model is usually used as a benchmark for other approaches.

$$x_{t+n} = x_t \tag{5}$$

3.2. ARMA

An Auto Regressive (AR) model $AR(p)$ is defined as:

$$x_t = \sum_{i=1}^{p} \phi_i x_{t-i} + w_t \tag{6}$$

where w_t is a Gaussian white noise with 0 mean and σ_w^2 variance.

$$w_t \sim \mathcal{WN}(0, \sigma_w^2)$$

Additional requirements are: (i) x_t is a stationary time series (ii) x_t is normally distributed and (iii) x_t should be normalized to 0 mean.

The AR model is to linearly correspond the time series with the historical data directly. An extension is to introduce some random variables which are called Moving Average (MA) to the model. $ARMA(p, q)$ is defined as:

$$x_t = \sum_{i=1}^{p} \phi_i x_{t-i} + w_t + \sum_{j=1}^{q} \theta_j w_{t-j} \tag{7}$$

where

$$w_t, w_{t-j} \sim \mathcal{WN}(0, \sigma_w^2)$$

3.2.1. ARIMA Modeling

One of the requirements of ARMA is that the time series should be stationary. However, most wind speed time series are non-stationary. One simple way to convert non-stationary time series to stationary one is to differencing the time series data:

$$\nabla^d x = (x_t - x_{t-1}) - (x_{t-1} - x_{t-2}) - \cdots - (x_{t-d+1} - x_{t-d}) \tag{8}$$

An ARIMA model $ARIMA(p, d, q)$ is defined as:

$$\nabla^d x_t = \sum_{i=1}^{p} \phi_i \nabla^d x_{t-i} + w_t + \sum_{j=1}^{q} \theta_j w_{t-j} \tag{9}$$

ARIMA can represent AR, MA or ARMA by setting the respective orders q, p or d to zero.

The way to determine p, d and q is by inspecting the histogram, ACF and PACF of the (differenced) time series. d is determined by trial and error, the best d is determined when the histogram of the differenced time series is normally distributed and the ACF of the differenced time series is cut-off rapidly. p and q are determined according to Table 1 [16].

Table 1. Model Selection of ARMA

		PACF	
		Exp. Decay	Cut-off at Lag p
ACF	Exp. Decay	-	ARMA(p,0)
	Cut-off at Lag q	ARMA(0,q)	ARMA (p,q)

3.3. ARIMA Parameter Estimation

Once p, d and q are determined, $\phi_1 \cdots \phi_p$ and $\theta_1 \cdots \theta_q$ can be determined by least square estimation or Yule-Walker estimation and Durbin's method.

Yule-Walker estimation [7] makes use of ACF to form a least square estimation problem. The $AR(p)$ equation can be re-written as:

$$x_i x_{i-p} = \sum_{j=1}^{p} \phi_j x_{i-p} x_{i-j} + x_{i-p} w_i \tag{10}$$

Taking expectations on both sides yields:

$$\mathrm{E}[x_i\, x_{i-p}] = \sum_{j=1}^{p} \phi_j\, \mathrm{E}[x_{i-p}\, x_{i-j}] \tag{11}$$

Thus:

$$\rho_p = \sum_{i=1}^{p} \phi_j \rho_{j-p} \tag{12}$$

or in matrix form:

$$\begin{bmatrix} \rho_1 \\ \rho_2 \\ \cdots \\ \rho_n \end{bmatrix} = \begin{bmatrix} 1 & \rho_1 & \cdots & \rho_{p-1} \\ \rho_1 & 1 & \cdots & \rho_{p-2} \\ \cdots & \cdots & \cdots & \cdots \\ \rho_{p-1} & \rho_{p-2} & \cdots & 1 \end{bmatrix} \begin{bmatrix} \phi_1 \\ \phi_2 \\ \cdots \\ \phi_p \end{bmatrix} \tag{13}$$

This is a conventional $\mathbf{P} = \mathbf{A}\Phi$ least square problem and $\hat{\Phi}$ can be estimated by $\hat{\Phi} = [\mathbf{A}^T\mathbf{A}]^{-1}\mathbf{A}^T\mathbf{P}$.

For $MA(q)$, Durbin's method [13] is usually used. Durbin's method is a two stage least square method, the first stage is to convert the $MA(q)$ to an $AR(p)$ problem with $p = 2q$ and the second stage is to estimate θ from the previously estimated results.

The $MA(q)$ can be fitted with $AR(p)$ as:

$$\hat{w}_t = y_t + \sum_{i=1}^{p} \phi_i y_{t-i} \tag{14}$$

and then

$$y_t - \hat{w}_t = \sum_{j=1}^{q} \theta_j w_{t-j} \tag{15}$$

Both of them can be solved by least square method as mentioned before.

3.4. ARIMA Forecasting

Forecasting is to predict future values x_{n+m} based on the data collected to the present $x_n, x_{n-1}, \cdots, x_1$. The forecasting is based on minimum mean square error, which is:

$$\hat{y}_S = x^n_{n+m} = E[x_{n+m}|x_n, x_{n-1}, \cdots, x_1] \qquad (16)$$

The forecasting process is rather complex and further reading is available at [4].

Some error measurements are usually used to evaluate the performance of the prediction. They are:

- Mean Absolute Error: $\text{MAE} = \text{E}[|\hat{y}_S - y_S|]$
- Root Mean Square Error: $\text{RMSE} = \sqrt{\text{E}[(\hat{y}_S - y_S)^2]}$
- Mean Absolute Percentage Error: $\text{MAPE} = \text{E}[|\frac{\hat{y}_S - y_S}{\text{E}[y_S]}|] \times 100\%$

4. Learning-based Approaches

Learning-based approaches discussed in this chapter consist of SVM and ELM. Both SVM and ELM are applicable to classification and regression. This section discusses the regression procedure of SVM and ELM.

4.1. Support Vector Machine Regression

SVM is a machine learning tool for classification as well as regression [6]. As a kernel machine, SVM can be implemented in a non-linear attribute space but with a linear classification hyperplane by appropriately choosing a kernel function. The kernel function helps SVM to perform well for non-linear datasets. However, choosing a kernel function and the parameters of the kernel function usually undergoes an exhaustive search process [10]. For SVM classification as shown in Figure 4(a), the data belonging to different classes are differentiated with either round shape or square shape. A hyperplane (solid line) and two soft margins (dashed line) separate the data by maximum separation and minimum misclassification error. For SVM regression as shown in Figure 4(b), a hyperplane (dashed line) is constructed to fit the trend of the time series data in the hyperspace (round shape in-between the dashed region) while allowing some tolerance for mistakes (round shape outside the dashed region). Notice that for both SVM classification and SVM regression, the data are mapped into hyperspace before processing and therefore the straight lines in Figure 4 are for illustration only.

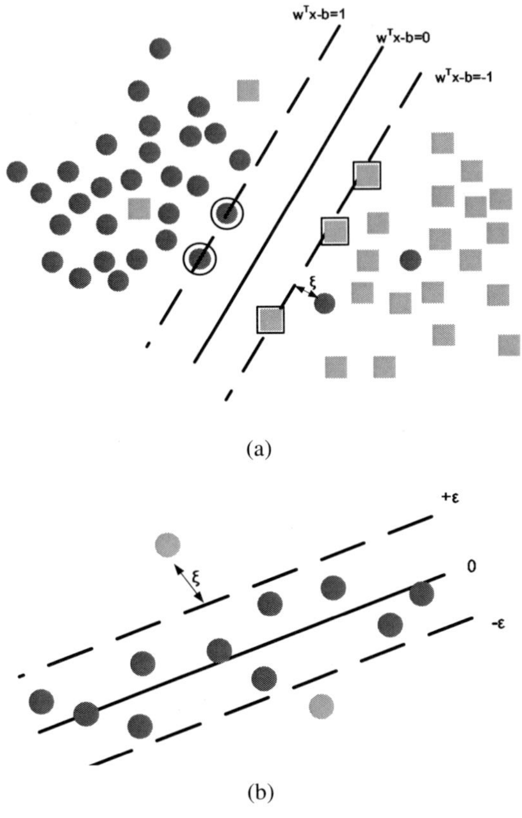

Figure 4. Visual Comparison of SVM (a) Classification and (b) Regression.

The objective of SVM regression is to find a maximally flatten hyperplane $f(x) = w^T x + b$ that has at most ϵ tolerance and minimal slack $\xi^{(*)}$ from the actually obtained targets y_i. The formal Lagrange Primal problem is defined as [6]:

$$\min \frac{1}{2}||w||^2 + C\sum_i(\xi_i + \xi_i^*) \tag{17}$$

subject to:

$$\begin{aligned} y_i - w^T x_i - b &\leq \epsilon + \xi_i \\ w^T x_i + b - y_i &\leq \epsilon + \xi_i^* \\ \xi_i^{(*)} &\geq 0 \end{aligned}$$

where C is the cost factor for trade-off between flatness and tolerance. $\xi_i^{(*)}$ is the slack variable.

The Lagrange Primal problem can be solved by the Lagrange Multiplier Method:

$$\begin{aligned} L :=& \frac{1}{2}||w||^2 + C\sum_i(\xi_i + \xi_i^*) - \sum_i(\eta_i\xi_i + \eta_i^*\xi_i^*) \\ &- \sum_i \alpha_i(\epsilon + \xi_i - y_i + w^T x_i + b) - \sum_i \alpha_i^*(\epsilon + \xi_i^* + y_i - w^T x_i - b) \end{aligned} \tag{18}$$

where $\alpha_i^{(*)}, \eta_i^{(*)} \geq 0$

$$\begin{aligned} \frac{\partial L}{\partial b} &= 0 \Rightarrow \sum_i(\alpha_i^* - \alpha_i) = 0 \\ \frac{\partial L}{\partial w} &= 0 \Rightarrow w = \sum_i(\alpha_i - \alpha_i^*)x_i \\ \frac{\partial L}{\partial \xi_i^{(*)}} &= 0 \Rightarrow C = \alpha_i^{(*)} - \eta_i^{(*)} \end{aligned}$$

So a Lagrange Dual problem is derived:

$$\max \sum_i y_i(\alpha_i - \alpha_i^*) - \epsilon\sum_i(\alpha_i + \alpha_i^*) - \frac{1}{2}\sum_i\sum_j(\alpha_i - \alpha_i^*)(\alpha_j - \alpha_j^*)x_i^T x_j \tag{19}$$

subject to:

$$\sum_i (\alpha_i - \alpha_i^*) = 0$$
$$0 \leq \alpha^{(*)} \leq C$$

The regression function $f(x) = w^T x + b$ can be re-written as:

$$f(x) = \sum_i (\alpha_i - \alpha_i^*) x_i^T x + b \tag{20}$$

The feature mapping is applied to the term $x_i^T x$. A kernel function $K(x_i, x_j)$ is used to map x_i to $\phi(x_i)$ with an implicit mapping function $\phi(\cdot)$ which is not explicitly evaluated. Kernel functions enables SVM regression to operate in a non-linear space, which is advantageous over statistical approaches. Some well known kernel functions are:

$$\text{Polynomial Kernel: } K(x_i, x_j) = (\gamma x_i^T x_j)^d$$
$$\text{RBF Kernel: } K(x_i, x_j) = e^{\gamma |x_i - x_j|^2}$$
$$\text{Sigmoid Kernel: } K(x_i, x_j) = \tanh(\gamma x_i^T x_j)$$

4.2. Extreme Learning Machine Regression

ELM is a SLFN universal learning machine as shown in Figure 5. It is suitable for both classification and regression without major variation. ELM is advantageous over conventional neural network because the hidden layer in ELM need not be iteratively trained, thus can be computed faster.

The regression function of ELM can be represented as:

$$f(x) = \mathbf{h}(x)\beta \tag{21}$$

where $x \in R^d$ and $\mathbf{h}(x) \in R^L$, $\mathbf{h}(\cdot)$ is a feature mapping function (kernel), β is an output weight vector.

The training process of ELM is to minimize the training error as well as the norm of the output weight. It can be represented as a Lagrange Primal problem:

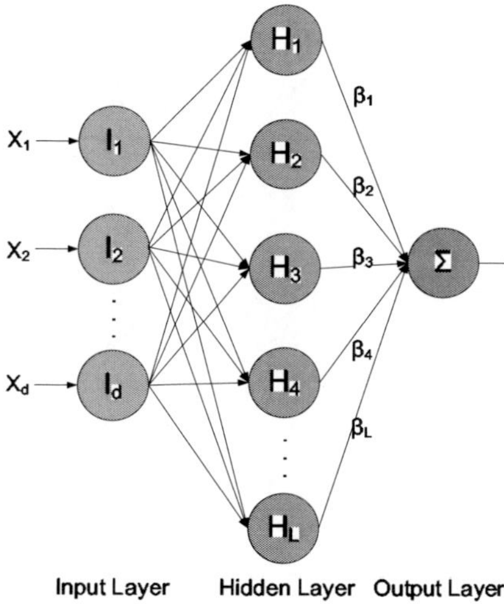

Figure 5. Schematic Diagram of ELM.

$$\min \frac{1}{2}||\beta||^2 + \frac{1}{2}C\sum_i \xi_i^2 \tag{22}$$

subject to:

$$y_i - \mathbf{h}(x_i)\beta = \xi_i$$

where:

$$\beta = \begin{cases} (\mathbf{H}^T\mathbf{H})^{-1}\mathbf{H}^T\mathbf{y}_T & \mathbf{H}^T\mathbf{H} \text{ is nonsingular,} \\ \mathbf{H}(\mathbf{H}\mathbf{H}^T)^{-1}\mathbf{y}_T & \mathbf{H}\mathbf{H}^T \text{ is nonsingular.} \end{cases}$$

and

$$\mathbf{H} = [\mathbf{h}(x_1)\ldots\mathbf{h}(x_n)]$$

The Lagrange Primal problem can be solved by the Lagrange Multiplier method:

$$L := \frac{1}{2}||\beta||^2 + \frac{1}{2}C\sum_i \xi_i^2 - \sum_i \alpha_i(\mathbf{h}(x_i)\beta - y_i + \xi_i) \tag{23}$$

$$\frac{\partial L}{\partial \beta} = 0 \Rightarrow \beta = \sum_i \alpha_i \mathbf{h}(x_i)^T = \mathbf{H}^T \alpha$$
$$\frac{\partial L}{\partial \xi_i} = 0 \Rightarrow \alpha_i = C\xi_i$$
$$\frac{\partial L}{\partial \alpha_i} = 0 \Rightarrow \mathbf{h}(x_i)\beta - y_i + \xi_i = 0$$

Finally, the regression function can be rewritten as:

$$f(x) = \mathbf{h}(x)\beta = \mathbf{h}(x)\mathbf{H}^T(\frac{\mathbf{I}}{C} + \mathbf{H}\mathbf{H}^T)^{-1})\mathbf{y}_T \quad (24)$$

4.3. Training and Testing

For learning-based forecasting, two phases are involved: training and testing/forecast The training phase is to determine the best regression function based on the known input data and corresponding output data. The testing phase is to use the trained regression function and the input data to predict the output data. A flow chart of the two phases (taking SVM as an example, ELM training and testing is the same) is shown in Figure 6.

The time series data needs to be scaled into the interval either $[-1, 1]$ or $[0, 1]$ so that SVM or ELM regression can perform well.

The scaled time series data also needs to be re-arranged before training and testing. The original wind speed time series data is in the one-dimensional vector format:

$$\{x_0, x_1, \ldots, x_{n-1}, x_n, \ldots, x_{n+m}\}$$

The re-arranged training and testing data are n-dimensional matrices:

$$\begin{bmatrix} x_{-n} & x_{-n+1} & \cdots & x_0 & | & x_m \\ x_{-n+1} & x_{-n+2} & \cdots & x_1 & | & x_{m+1} \\ & \ddots & & & \vdots & \\ x_1 & x_2 & \cdots & x_n & | & x_{m+n} \end{bmatrix} \quad (25)$$

where n is the number of historical data visited and m is the time horizon of prediction.

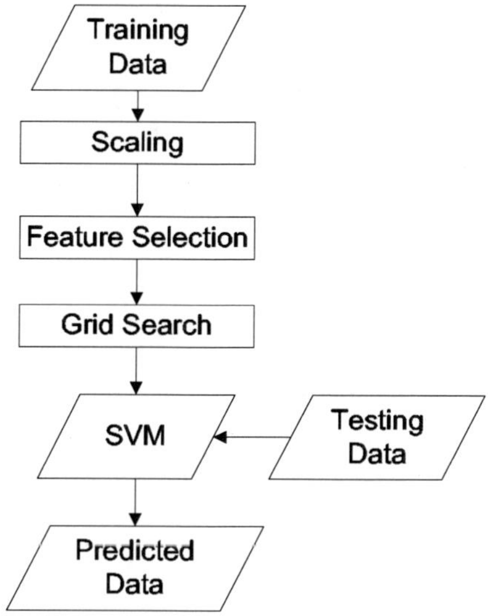

Figure 6. Flow Chart of Learning-based Forecasting.

The grid search method is used to find the parameters associated with the best performing regression model. Both SVM and ELM need grid search to tune the parameters. For SVM regression, penalty factor C and kernel parameters such as γ and d need to be tuned whereas for ELM, penalty factor C and hidden layer number L need to be tuned. The grid search uses n-fold CV to minimize over-fitting.

In the testing phase, the scaled test data are predicted with the trained model and the output is the predicted data.

5. Results

The wind speed dataset is retrieved from NDBC [2] site 41001 located at (34.561N, 72.631W) during year 2011. The resolution is 10 minutes average. There was no data record from mid October to early November. The wind speed time series of year 2011 is plotted in Figure 1. Detailed analysis is carried out for each month. For each month, the first 70% data points are used as training data and the remaining 30% data points are used as testing data. The starting and ending time of each monthly training and testing data are tabulated in Tables 2 and 3.

For ARIMA, the software used is R 2.15.1 [12] and for SVM and ELM, the software used is Matlab R2011b [1]. The SVM package in Matlab is libsvm 3.11 [5] and the ELM code in Matlab is retrieved from [9].

Table 2. Training Dataset Description

Month	Start	End	Samples
Dec	10-12-01 00:00	10-12-22 16:00	3103
Jan	11-01-01 00:00	11-01-22 16:10	3116
Feb	11-02-01 00:00	11-02-20 14:50	2814
Mar	11-03-01 00:00	11-03-22 16:50	3120
Apr	11-04-01 00:00	11-04-22 00:00	3019
May	11-05-01 00:00	11-05-22 16:50	3120
Jun	11-06-01 00:00	11-06-21 23:50	3024
Jul	11-07-01 00:00	11-07-22 15:50	3120
Aug	11-08-01 00:00	11-08-22 17:00	3103
Sep	11-09-01 00:00	11-09-22 00:00	3019
Oct	11-10-01 00:00	11-10-09 06:50	1188
Nov	11-11-01 03:00	11-11-23 03:10	2402

Conventional ARIMA model fitting requires ACF and PACF plotting and maximum likelihood estimation. With the help of software, the AR (p), differencing (d) and MA (q) orders and the AR (ϕ) and MA (θ) coefficients are determined by 'auto.arima()' function in R 2.15.1. The coefficients are tabulated in Table 4. For January data, the 'auto.arima()' function returns an ARMA model instead of ARIMA model. For the remaining monthly data, one-step differencing ($d = 1$) is determined by the 'auto.arima()' function.

The fitted ARIMA models are tested on the testing data with 6-step ahead forecast (1 hour ahead), 12-step ahead forecast (2 hours ahead) and 24-step ahead forecast (4 hours ahead).

The training and testing of monthly wind speed time series data with SVM and ELM are implemented in Matlab. The forecasting results are compared with the persistent model and the ARIMA model and three error measurements are

Table 3. Testing Dataset Description

Month	Start	End	Samples
Dec	10-12-22 16:10	10-12-31 23:50	1331
Jan	11-01-22 16:20	11-01-31 23:50	1336
Feb	11-02-20 15:00	11-02-28 23:50	1206
Mar	11-03-22 17:00	11-03-31 23:50	1338
Apr	11-04-22 00:10	11-04-30 23:50	1295
May	11-05-22 17:00	11-05-31 23:50	1338
Jun	11-06-22 00:00	11-06-30 23:50	1296
Jul	11-07-22 16:00	11-07-31 23:50	1338
Aug	11-08-22 17:10	11-08-31 23:50	1331
Sep	11-09-22 00:10	11-09-30 23:50	1295
Oct	11-10-09 07:00	11-10-31 10:50	510
Nov	11-11-23 03:20	11-11-30 06:50	1030

Table 4. Parameter Estimation of ARIMA model for Monthly Training Data

Month	Orders	ϕ_1	ϕ_2	ϕ_3	ϕ_4	ϕ_5	θ_1	θ_2	θ_3	θ_4
Dec	(3,1,4)	-0.89	-0.97	-0.41	-	-	0.49	0.48	-0.10	-0.30
Jan	(1,0,2)	0.99	-	-	-	-	-0.27	-0.05	-	-
Feb	(1,1,1)	0.63	-	-	-	-	-0.70	-	-	-
Mar	(2,1,0)	-0.16	-0.10	-	-	-	-	-	-	-
Apr	(5,1,3)	-0.03	0.22	0.14	0.10	0.03	-0.04	-0.28	-0.16	-
May	(1,1,2)	0.81	-	-	-	-	-0.93	0.03	-	-
Jun	(0,1,1)	-	-	-	-	-	-0.16	-	-	-
Jul	(1,1,1)	0.78	-	-	-	-	-0.85	-	-	-
Aug	(3,1,3)	-0.34	-0.03	0.37	-	-	0.29	-0.13	-0.43	-
Sep	(1,1,4)	0.60	-	-	-	-	-0.57	-0.09	-0.04	-0.06
Oct	(1,1,1)	0.23	-	-	-	-	-0.44	-	-	-
Nov	(4,1,4)	-0.15	-0.18	-0.32	0.23	-	-0.12	0.04	0.27	-0.39

used: RMSE, MAE and MAPE. The results are tabulated in Tables 5, 6 and 7. The lowest errors are highlighted with bold font.

As shown in the tables, the errors increase with increasing the forecasting time horizon regardless of the error measurement types. For example, the ARIMA model's RMSE of 1-hour, 2-hour and 4-hour ahead forecasting on April data are 0.0516, 0.0648 and 0.0725 respectively. The ELM model's

Table 5. Error Measurement of Wind Speed Forecast, 1 Hour Ahead

Month	RMSE	MAE	MAPE
Persistent			
Dec	**0.0455**	0.0335	6.9185
Jan	0.0650	0.0474	10.3348
Feb	0.0485	0.0380	8.9933
Mar	0.0679	0.0459	**13.4127**
Apr	0.0507	0.0343	9.3678
May	0.0389	**0.0273**	11.5451
Jun	0.0551	0.0382	11.8370
Jul	0.0407	0.0276	8.5402
Aug	0.0624	0.0429	11.7974
Sep	0.0615	0.0385	**10.4510**
Oct	0.0428	0.0280	7.8624
Nov	0.0555	0.0402	11.3483
ARIMA			
Dec	0.0469	**0.0330**	**6.8250**
Jan	0.0637	0.0464	**10.1050**
Feb	0.0483	0.0380	8.9388
Mar	**0.0673**	**0.0457**	13.4211
Apr	0.0516	0.0347	9.4078
May	0.0389	0.0273	11.6078
Jun	0.0543	0.0377	11.8965
Jul	0.0407	0.0275	**8.4598**
Aug	**0.0613**	**0.0426**	**11.7405**
Sep	0.0612	**0.0384**	10.4929
Oct	**0.0420**	**0.0273**	**7.6023**
Nov	0.0553	**0.0394**	**11.1266**
ELM			
Dec	0.0478	0.0370	7.6276
Jan	0.0644	0.0466	10.1633
Feb	**0.0473**	**0.0371**	**8.7882**
Mar	0.0682	0.0464	13.5561
Apr	**0.0505**	**0.0341**	**9.3174**
May	0.0389	0.0276	11.7045
Jun	0.0541	0.0372	**11.5247**
Jul	**0.0399**	**0.0275**	8.5261
Aug	0.0700	0.0503	13.8337
Sep	**0.0606**	0.0388	10.5546
Oct	0.0431	0.0291	8.1540
Nov	**0.0544**	0.0400	11.2916
SVM			
Dec	0.0544	0.0418	8.4405
Jan	**0.0632**	**0.0461**	10.1279
Feb	0.0482	0.0381	9.0434
Mar	0.0685	0.0470	13.5026
Apr	0.0509	0.0348	9.4674
May	**0.0380**	0.0274	**11.3851**
Jun	**0.0534**	**0.0370**	11.6757
Jul	0.0400	0.0278	8.6085
Aug	0.0694	0.0506	14.5256
Sep	0.0611	0.0395	10.9435
Oct	0.0450	0.0303	8.5134
Nov	0.0554	0.0407	11.4723

Table 6. Error Measurement of Wind Speed Forecast, 2 Hours Ahead

Month	RMSE	MAE	MAPE
Persistent			
Dec	**0.0646**	**0.0470**	**9.6994**
Jan	0.0882	0.0681	14.8764
Feb	0.0697	0.0552	13.0886
Mar	0.0882	**0.0645**	18.8604
Apr	0.0633	**0.0431**	**11.7890**
May	**0.0479**	**0.0364**	15.4311
Jun	0.0694	0.0505	15.5784
Jul	0.0563	0.0401	12.4202
Aug	0.0769	0.0556	15.2988
Sep	0.0827	0.0530	14.3865
Oct	0.0536	0.0359	10.0956
Nov	0.0736	0.0554	15.6455
ARIMA			
Dec	0.0661	0.0472	9.7770
Jan	0.0861	0.0662	14.4114
Feb	0.0692	0.0549	12.9318
Mar	**0.0877**	0.0647	18.9811
Apr	0.0648	0.0437	11.8607
May	0.0480	0.0365	15.5390
Jun	0.0686	0.0500	15.7753
Jul	0.0564	0.0400	12.2972
Aug	**0.0753**	**0.0551**	**15.1669**
Sep	0.0826	0.0532	14.5323
Oct	0.0523	**0.0352**	**9.8232**
Nov	0.0728	0.0549	15.5032
ELM			
Dec	0.0682	0.0540	11.1311
Jan	0.0846	0.0646	**14.0940**
Feb	**0.0656**	**0.0526**	**12.4630**
Mar	0.0898	0.0658	19.2276
Apr	0.0653	0.0438	11.9784
May	0.0488	0.0376	15.9049
Jun	0.0673	**0.0490**	**15.1224**
Jul	0.0539	0.0389	12.0575
Aug	0.1013	0.0728	20.0261
Sep	**0.0796**	**0.0527**	**14.3209**
Oct	**0.0517**	0.0373	10.4824
Nov	0.0730	0.0552	15.5811
SVM			
Dec	0.0802	0.0630	12.4860
Jan	**0.0836**	**0.0640**	14.1574
Feb	0.0663	0.0536	12.7259
Mar	0.0878	0.0656	**18.5056**
Apr	**0.0633**	0.0435	11.8169
May	0.0482	0.0374	**15.2667**
Jun	**0.0670**	0.0493	15.8322
Jul	**0.0533**	**0.0386**	**11.9566**
Aug	0.0950	0.0700	20.7741
Sep	0.0797	0.0529	14.8574
Oct	0.0532	0.0379	10.6775
Nov	**0.0725**	**0.0548**	**15.3976**

Table 7. Error Measurement of Wind Speed Forecast, 4 Hours Ahead

Month	RMSE	MAE	MAPE
Persistent			
Dec	**0.1019**	**0.0741**	**15.2589**
Jan	0.1224	0.0963	21.1138
Feb	0.1081	0.0864	20.5129
Mar	0.1086	0.0847	24.7253
Apr	0.0715	0.0531	14.5968
May	0.0638	0.0501	21.2103
Jun	0.0855	0.0667	**20.4064**
Jul	0.0736	0.0551	17.0990
Aug	0.0946	0.0747	20.4850
Sep	0.1082	0.0723	19.6456
Oct	0.0771	0.0544	15.3170
Nov	0.0959	**0.0758**	**21.4764**
ARIMA			
Dec	0.1033	0.0745	15.4115
Jan	0.1167	0.0921	20.0529
Feb	0.1077	0.0861	20.2695
Mar	0.1074	**0.0837**	24.5622
Apr	0.0725	0.0536	14.5320
May	**0.0636**	**0.0501**	21.3131
Jun	0.0847	0.0661	20.8557
Jul	0.0734	0.0551	16.9565
Aug	**0.0935**	**0.0743**	**20.4568**
Sep	0.1078	0.0724	19.7808
Oct	**0.0757**	**0.0532**	**14.8367**
Nov	0.0962	0.0768	21.6635
ELM			
Dec	0.1069	0.0860	17.7114
Jan	0.1200	0.0941	20.6233
Feb	0.0990	**0.0788**	**18.7057**
Mar	0.1130	0.0890	25.9810
Apr	0.0712	0.0521	14.3183
May	0.0679	0.0541	22.9155
Jun	0.0927	0.0696	21.3019
Jul	0.0676	**0.0512**	**15.8832**
Aug	0.1138	0.0861	23.5971
Sep	0.1002	0.0711	**19.3081**
Oct	0.0773	0.0581	16.3647
Nov	0.0961	0.0763	21.6206
SVM			
Dec	0.1225	0.0977	18.9301
Jan	**0.1130**	**0.0885**	**19.8555**
Feb	**0.0983**	0.0791	18.7876
Mar	**0.1065**	0.0845	**23.2493**
Apr	**0.0683**	**0.0520**	**14.0325**
May	0.0665	0.0532	**20.9716**
Jun	**0.0841**	**0.0642**	21.1096
Jul	**0.0675**	0.0515	16.0280
Aug	0.1271	0.0926	28.6790
Sep	**0.0990**	**0.0706**	20.2414
Oct	0.0773	0.0570	16.0929
Nov	**0.0958**	0.0769	21.6128

MAPE of 1-hour, 2-hour and 4-hour ahead forecasting on April data are 9.3174%, 11.9784% and 14.3183% respectively.

In general, ARIMA, SVM and ELM outperform the persistent model for the three tested time horizons. For 1-hour ahead forecasting, only RMSE of December, MAPE of March and September, and MAE of May have the persistent model outperformed the rest. For 2-hour ahead forecasting, only RMSE of December and May, MAE of December, March, April and May, and MAPE of December and April have the persistent model outperforming the rest. For 4-hour ahead forecasting, only RMSE of December, MAE of December and November, and MAPE of December and June and November have the persistent model outperforming the rest.

For shorter-term forecasting, ARIMA is generally more accurate than the learning-based approaches whereas for longer-term forecasting, ARIMA is outperformed by the learning-based approaches for most of the monthly wind time series data. For 1-hour a head forecasting, the RMSE, MAE and MAPE comparisons show that ARIMA outperforms the other two models on 4, 7 and 8 out of 12 months' data, respectively. For 4-hour a head forecasting, the RMSE, MAE and MAPE comparisons show reduced fractions (4, 5, 3 respectively) of ARIMA's out-performance.

The comparisons between ELM and SVM show that for shorter-term forecasting, ELM outperforms SVM but for longer-term forecasting, SVM outperforms ELM. For 1-hour ahead forecasting, the RSME, MAE and MAPE of ELM outperforming SVM are 8, 9 and 9 out of 12 months' data. For 2-hour ahead forecasting, the RSME, MAE and MAPE of ELM outperforming SVM reduce to 4, 5 and 7 out of 12 months' data. However, for 4-hour ahead forecasting, the RSME, MAE and MAPE of ELM outperforming SVM further reduce to 2, 5 and 5 out of 12 months' data.

Conclusion

This chapter has compared a statistical approach (ARIMA) with two learning-based approaches (SVM and ELM) on wind speed time series forecasting. The statistics of a wind speed time series have been analyzed. The mathematics and procedures of ARIMA, SVM and ELM have been elaborated. A historical off-shore wind speed data has been selected for forecasting. The results have shown that (i) the forecasting error increases with increasing the time horizon; (ii) ARIMA is generally more preferable for short-term forecasting. (iii) the forecasting accuracy of ELM is higher than SVM for shorter-term but lower for longer-term.

References

[1] Matlab r2011b, 2011.

[2] National data buoy center historical data, July 2012.

[3] J. B. Bremnes. A comparison of a few statistical models for making quantile wind power forecasts. *Wind Energy*, 9:3–11, 2006.

[4] Peter J. Brockwell and Richard A. Davis. *Introduction to Time Series and Forecasting*. NY:Springer, 2002.

[5] Chih-Chung Chang and Chih-Jen Lin. LIBSVM: A library for support vector machines. *ACM Transactions on Intelligent Systems and Technology*, 2:27:1–27:27, 2011. Software available at http://www.csie.ntu.edu.tw/ cjlin/libsvm.

[6] C. Cortes and V. Vapnik. Support-vector networks. *Machine Learning*, 20(3):273–297, 1995.

[7] Gidon Eshel. The Yule Walker equations for the AR coefficients. Technical report, University of South Carolina, 2010.

[8] David C. Hill, David McMillan, Keith R. W. Bell, and David Infield. Application of auto-regressive models to U.K. wind speed data for power system impact studies. *IEEE Transactions on Sustainable Energy*, 3(1):134–141, January 2012.

[9] G.-B. Huang, H. Zhou, X. Ding, and R. Zhang. Extreme learning machine for regression and multi-class classification. *IEEE Transactions on Systems, Man, and Cybernetics, Part B: Cybernetics*, PP(99):1–17, October 2011.

[10] Ruhai Lei, Xiaoxiao Kong, and Xuesong Wang. An ensemble SVM using entropy-based attribute selection. In *Chinese Control and Decision Conference*, pages 802 –805, May 2010.

[11] M. Lydia and S.S. Kumar. A comprehensive overview on wind power forecasting. In *2010 9th International Power and Energy Conference (IPEC 2010)*, pages 268 – 73, Piscataway, NJ, USA, 2010.

[12] R Core Team. *R: A Language and Environment for Statistical Computing*. R Foundation for Statistical Computing, Vienna, Austria, 2012. ISBN 3-900051-07-0.

[13] Niclas Sandgren and Petre Stoica. On moving average parameter estimation. Technical Report 2006-022, Department of Information Technology, Uppsala University, April 2006.

[14] Robert H. Shumway and David S. Stoffer. *Time Series Analysis and Its Applications, with R examples*. NY:Springer, 2011.

[15] S.S. Soman, H. Zareipour, O. Malik, and P. Mandal. A review of wind power and wind speed forecasting methods with different time horizons. In *North American Power Symposium (NAPS), 2010*, pages 1 –8, sept. 2010.

[16] Vincent Zoonekynd. Time series, January 2007.

In: Smart Microgrids
Editor: Hussein M. Khodr
ISBN:978-62808-975-2

Chapter 2

SHORT-TERM SCHEDULING OF MICROGRID WITH RENEWABLE SOURCES AND COMBINED HEAT AND POWER

Mohammad Moradi-Dalvand[1*], ***Behnam Mohammadi-Ivatloo*** [2†]
and Mohammad Ali Fotouhi Ghazvini [3‡]
[1] Electrical Engineering Department, Power and Water University of Technology, Tehran, Iran
[2] Faculty of Electrical and Computer Engineering, University of Tabriz, Tabriz, Iran
[3] GECAD, Knowledge Engineering and Decision Support Research Center, Polytechnic of Porto (IPP), Porto, Portugal

Abstract

This chapter studies the short-term scheduling problem of an isolated microgrid with combined heat and power (CHP) units and renewable energy resources such as wind and solar energies. The considered microgrid can be an isolated power system, an urban area or a large company with

[*]E-mail address: moradi.dalvand@gmail.com
[†]E-mail address: mohammadi@ieee.org
[‡]E-mail address: ma.fotouhi@gmail.com

heat and power demand in a real world system. The objective of the short-term scheduling is determining the optimal schedule of different power and heat generation units for serving the power and heat demands with minimum cost. The technical constraints of heat and power generators are considered in the proposed optimization framework. In this chapter, the dual dependency between generated heat and power of CHP units is considered and mathematical models of different CHP types are derived. The effect of uncertainty of wind and solar energy productions are considered using the implemented reserve requirement approach. In this approach a portion of forecasted load and renewable generation, i.e. wind and solar generation, is considered as the desirable system reserve. A battery storage is incorporated in the studied system model. Using energy storage systems increase the utilization of the renewable energy resources and the total system efficiency. The proposed model is formulated as a mixed integer nonlinear programming problem and solved using GAMS. An illustrative example is presented in order to demonstrate the applicability of the proposed method.

Keywords: Microgrid, Renewable Energy, Combined Heat and Power (CHP), Energy storage

Nomenclature

Indices:

t	Index for operation intervals running from 1 to T.
i	Index for power-only generation units running from 1 to N_{po}.
j	Index for CHP units running from 1 to N_{chp}.
k	Index for heat-only units running from 1 to N_{ho}.
l	Index for Photovoltaic units running from 1 to N_{pv}.
m	Index for wind power units running from 1 to N_w.

Parameters:

c_l^{pv}	Linear operation cost coefficient of PV unit l.
c_m^w	Linear operation cost coefficient of wind power unit m.
P_t^D	The electric power demand of the system at time t (MWh).
H_t^D	The heat demand of the system at time t (MWh-th).
$P_i^{po,max}(P_i^{po,min})$	Maximum (minimum) power output of i-th power-only unit (MW).
$P_l^{pv,max}(P_l^{pv,min})$	Maximum (minimum) power output of l-th PV unit (MW).
$P_m^{w,max}(P_m^{w,min})$	Maximum (minimum) power output of m-th wind power unit (MW).
$H_k^{ho,max}(H_k^{ho,min})$	Maximum (minimum) heat output of k-th heat-only unit (MW-th).
WS_t	The forecasted wind speed at time t (m/s).
V_k^{cutin}	The cut in speed of wind turbine k (m/s).
V_k^{cutout}	The cut out speed of wind turbine k (m/s).
V_k^{rated}	The rated speed of wind turbine k (m/s).
$P_k^{w,rated}$	The rated power of wind turbine k (MW).
$P_{kt}^{w,av}$	Forecasted generation of wind turbine k at time t (MWh).
x_t	Forecasted solar radiation at time t (W/m^2).
$P_{lt}^{pv,av}$	Forecasted solar power of PV l at time t (MWh).
$A_l^{pv}, B_l^{pv}, C_l^{pv}$	Generation power function coefficients of PV l.
$E^{max}(E^{min})$	The maximum (minimum) level of energy storage (MWh).
$P_{max}^{cha}(P_{min}^{cha})$	The maximum (minimum) charge rate of the storage (MW).

$P_{max}^{dch}(P_{min}^{dch})$	The maximum (minimum) discharge rate of the storage (MW).
$\eta_{cha}(\eta_{dch})$	The charging (discharging) efficiency of the storage (%).
$suc_i^{po}(sdc_i^{po})$	The start-up (shut-down) cost of power-only unit i (\$).
$suc_j^{chp}(sdc_j^{chp})$	The start-up (shut-down) cost of CHP unit j (\$).
$suc_k^{ho}(sdc_k^{ho})$	The start-up (shut-down) cost of heat-only unit k (\$).
$a_i^{ho}, b_i^{ho}, c_i^{ho}$	Quadratic cost coefficients of power-only unit i.
$a_k^{ho}, b_k^{ho}, c_k^{ho}$	Quadratic cost coefficients of heat-only unit k.
$a_j^{chp}, b_j^{chp}, c_j^{chp}$ $, d_j^{chp}, e_j^{chp}, f_j^{chp}$	Cost coefficients of CHP unit j.
M	A sufficient large number.
k_1, k_2	Weighting coefficients of total power demand and dispatched intermittent powers in determining reserve requirement.

Variables:

P_{it}^{po}	Power output of i-th power-only unit at time t (MWh).
P_{jt}^{chp}	Power output of j-th CHP unit at time t (MWh).
P_{lt}^{pv}	Power output of l-th PV unit at time t (MWh).
P_{mt}^{w}	Power output of m-th wind power unit at time t (MWh).
H_{jt}^{chp}	Heat output of j-th CHP unit at time t (MWh-th).
H_{kt}^{ho}	Heat output of k-th heat-only unit at time t (MWh-th).
R_t	System reserve requirement at time t (MW).

V_{it}^{po}	Binary decision variable for commitment status of power-only unit i at time t.
V_{jt}^{chp}	Binary decision variable for commitment status of CHP unit j at time t.
E_t^s	The amount of stored energy in storage at time t (MWh).
P_t^{cha}	The charging power of the storage unit during time t (MW).
P_t^{dch}	The discharging power of the storage unit during time t (MW).
$y_{it}^{po}(z_{it}^{po})$	Binary decision variable representing start-up (shut-down) status of power-only unit i at time t.
$y_{kt}^{ho}(z_{kt}^{ho})$	Binary decision variable representing start-up (shut-down) status of heat-only unit k at time t.
$y_{jt}^{chp}(z_{jt}^{chp})$	Binary decision variable representing start-up (shut-down) status of CHP unit j at time t.
$P_{jt}^{chp,max}(H_{jt}^{chp})$	Maximum power output of j-th CHP unit when its heat is equal to H_{jt}^{chp} (MW).

Functions:

$C_i(P_i^{po})$	Cost function of the i-th power-only unit.
$C_j(P_j^{chp}, H_j^{chp})$	Cost function of the j-th CHP unit.
$C_k(H_k^{ho})$	Cost function of the k-th heat-only unit.
$C_l(P_l^{pv})$	Cost function of the l-th PV unit.
$C_m(P_m^{w})$	Cost function of the m-th wind power unit.

1 Introduction

1.1 Motivation

Interconnection of small generators or energy storage units at low or medium voltage power systems is the common characteristic of most microgrids [1]. The concept of microgrids enhances the marketability of microsources and provides a reassuring perspective for the future of power systems with scarcity of fossil fuels [2]. A smart microgrid has distributed generation (DG) units mainly powered from renewable resources and energy storage systems with an intelligent infrastructure at the end points of the network for active energy flow management. The generating units with 100 MW generation capacity are generally known as DGs, and mainly are connected to the distribution level fairly close to the end use nodes [3]. From the viewpoint of the system operator, a microgrid is an controllable / curtaillable electrical load which can supply loads isolated from the grid when the system is under stress [1, 2].

Higher energy efficiency with significant energy saving is the result of simultaneous production of heat and electric power [4]. Combined heat and power generation improves energy conversion efficiency and therefore reduces need for more fossil fuels, which consequently leads to less environmental pollutants. CHPs can be designed according to the heat demand of the system with the power as a by-product, or vice versa, excess generated heat of a power plant can be used [4,5]. To determine the optimal schedule of power and heat units, it is necessary to consider the operational characteristics in the economic dispatch problem of the system. The short-term operational scheduling of CHP units in hybrid microgrids with significant share of DG units is highly demanding. It is a non-convex and nonlinear problem with a complex structure due to the dual dependency of heat and power production and their joint characteristics [4, 6]. In other words, the interdependence of variables in short-term CHP operational scheduling makes the problem more difficult to solve [6].

The network interconnected microgrid should be able to disconnect from the grid while the system is under the stress of an abnormal situation and it should be able to operate in the islanded mode [2]. DG units can be used to enhance the reliability of the system by providing backup generation for a microgrid during islanding operation mode [7]. They also can provide ancillary services (e.g., voltage support and reactive power control) for the microgrid during grid-connected operation [7].

1.2 Review of Related Works

Unit commitment and economic dispatch are the main attention in short-term CHP planning, mainly within the time horizons of one day and hourly time increments. An optimization model has been developed in [3] for economic dispatch of a system with CHP units and wind turbines. A Particle Swarm Optimization (PSO) algorithm with time varying acceleration coefficients has been proposed and implemented in [8] for several test systems to solve the CHP economic dispatch problem. The coefficients vary adaptively during iterations to improve algorithm speed and to obtain better solutions [8]. A dynamic regrouping algorithm based on sequential dynamic programming has been proposed in [4] for short-term CHP production planning with relaxed ON/OFF states of the CHP units and sequential commitment of them in small groups using dynamic programming. The effect of heat accumulators and storage units in profitability of CHP units is investigated in [9]. Moreover, the proposed model in [9] determines the optimal storage capacity for different price scenarios.

The trigeneration module takes CHPs one step ahead by utilizing the waste heat to produce cooling [10]. In other words, the primary energy sources are converted to electric power, heat and cooling in trigeneration. Medium-term and long-term trigeneration planning with energy storage systems in a deregulated power market is investigated in [10]. The optimization model for a microgrid with multiple DGs, such as solar, wind and energy storage systems in has been presented in [1]. PSO algorithm has been implemented in [1] to minimize migrogrid cost in different market policies. The microgrid cost in this paper is annualized cost which is the summation of annual capital, replacement, operation and maintenance costs.

Storage technologies greatly facilitate the secure and reliable operation of the microgrids with significant share of renewable units. They are also expected to offer services such as spinning reserve for the microgrid [2]. The short-term operation of a conventional gas-fired generation company that possesses wind units and compressed air energy storage (CAES) is the purpose of the proposed analytical framework in [11]. The proposed mixed integer nonlinear programming formulation consists in the constraints of ramp rate, capacity, and minimum on/off time for the conventional units [11]. Short-term generation and maintenance outage scheduling of a microgrid with several energy storage units has been modeled modeled in [12], and it has been noted that the storage units are more profitable when the operator of the microgrid plans to increase the share of intermittent renewable units for supplying the demand. The system operator of a network with high penetration of wind generation should increases the reserve requirement by growth of wind power generation operate system securely [13]. The operator ensures that the reserve is available to cover the unanticipated changes of wind and solar output in addition to transmission line or generator outages. In this chapter the optimal operation of microgrid with renewable energy sources and CHP units is discussed and formulated.

1.3 Chapter Organization

The remainder of the chapter is organized as follows. The proposed problem formulation is presented in Section 2. A comprehensive illustrative example with different case studies is provided in Section 3. Finally the concluding remarks are provided in Section 3.2.

2 Problem Formulation

2.1 Objective Function

The objective of the problem is to minimize the total cost of serving power and heat demand s during the operation horizon. The power demand can be provided by CHP, power-only, wind and solar generation units. The heat demand is provided by CHP units and boilers (heat-only units). The objective function can be mathematically expressed as follows:

$$\begin{aligned}\min \sum_{t=1}^{T}[\sum_{i=1}^{N_{po}} C_i(P_{it}^{po}) + suc_i^{po} \times max(0, V_{it}^{po} - V_{it-1}^{po}) + sdc_i^{po} \times max(0, V_{it-1}^{po} - V_{it}^{po})\\ + \sum_{j=1}^{N_{chp}} C_j(P_{jt}^{chp}, H_{jt}^{chp}) + suc_j^{chp} \times max(0, V_{jt}^{chp} - V_{jt-1}^{chp}) + sdc_j^{chp} \times max(0, V_{jt-1}^{chp} - V_{jt}^{chp})\\ + \sum_{k=1}^{N_{ho}} C_k(H_{kt}^{ho}) + suc_k^{ho} \times max(0, V_{kt}^{ho} - V_{kt-1}^{ho}) + sdc_k^{ho} \times max(0, V_{kt-1}^{ho} - V_{kt}^{ho})\\ + \sum_{l=1}^{N_{pv}} C_l(P_{lt}^{pv}) + \sum_{m=1}^{N_w} C_m(P_{mt}^{w})] \ \$ \end{aligned} \quad (1)$$

Two binary variables can be used for start up and shut down conditions of each unit at each time interval as (2) and (3).

$$\begin{aligned} y_{it}^{po} &\geq V_{it}^{po} - V_{it-1}^{po} \\ z_{it}^{po} &\geq V_{it-1}^{po} - V_{it}^{po} \\ y_{jt}^{chp} &\geq V_{jt}^{chp} - V_{jt-1}^{chp} \\ z_{jt}^{chp} &\geq V_{jt-1}^{chp} - V_{jt}^{chp} \\ y_{kt}^{ho} &\geq V_{kt}^{ho} - V_{kt-1}^{ho} \\ z_{kt}^{ho} &\geq V_{kt-1}^{ho} - V_{kt}^{ho} \end{aligned} \tag{2}$$

$$\begin{aligned} y_{it}^{po} &\in \{0,1\} \\ z_{it}^{po} &\in \{0,1\} \\ y_{jt}^{chp} &\in \{0,1\} \\ z_{jt}^{chp} &\in \{0,1\} \\ y_{kt}^{ho} &\in \{0,1\} \\ z_{kt}^{ho} &\in \{0,1\} \end{aligned} \tag{3}$$

Now we can rewrite the objective function as (4).

$$\begin{aligned} \min \sum_{t=1}^{T} [\sum_{i=1}^{N_{po}} C_i(P_{it}^{po}) + suc_i^{po} \times y_{it}^{po} + sdc_i^{po} \times z_{it}^{po} \\ + \sum_{j=1}^{N_{chp}} C_j(P_{jt}^{chp}, H_{jt}^{chp}) + suc_j^{chp} \times y_{it}^{chp} + sdc_j^{chp} \times z_{it}^{chp} \\ + \sum_{k=1}^{N_{ho}} C_k(H_{kt}^{ho}) + suc_k^{ho} \times y_{kt}^{ho} + sdc_k^{ho} \times z_{kt}^{ho} \\ + \sum_{l=1}^{N_{pv}} C_l(P_{lt}^{pv}) + \sum_{m=1}^{N_w} C_m(P_{mt}^{w})] \ \$ \end{aligned} \tag{4}$$

The production cost of the power-only, CHP and heat-only units can be expressed as follows [8]:

$$C_i(P_{it}^{po}) = a_i^{po}(P_{it}^{po})^2 + b_i^{po}P_{it}^{po} + c_i^{po} \quad \$/h \tag{5}$$

$$C_j(P_{jt}^{chp}, H_{jt}^{chp}) = a_j^{chp}(P_{jt}^{chp})^2 + b_j^{chp}P_{jt}^{chp} + c_j^{chp} \tag{6}$$
$$+d_j^{chp}(H_{jt}^{chp})^2 + e_j^{chp}H_{jt}^{chp} + f_j^{chp}H_{jt}^{chp}P_{jt}^{chp} \quad \$/h$$

$$C_k(H_{kt}^{ho}) = a_k^{ho}(H_{kt}^{ho})^2 + b_k^{ho}H_{kt}^{ho} + c_k^{ho} \quad \$/h \tag{7}$$

The linear operation cost function is the popular and standardized representation for wind and PV operation costs.

$$C_l(P_{lt}^{pv}) = c_l^{pv}(P_{lt}^{pv}) \quad \$/h \tag{8}$$

$$C_m(P_{mt}^{w}) = c_m^{w}(P_{mt}^{w}) \quad \$/h \tag{9}$$

It should be mentioned that the operation cost of the wind and PV units are low and are often neglected in literature [11, 14].

2.2 Constraints

The following constraints should be considered in minimizing the objective function expressed in (1).

1. Power production and demand balance

$$\sum_{i=1}^{N_{po}} P_{it}^{po} + \sum_{j=1}^{N_{chp}} P_{jt}^{chp} + \sum_{l=1}^{N_{pv}} P_{lt}^{pv} + \sum_{m=1}^{N_w} P_{mt}^{w} + P_t^{dch} = P_t^D + P_t^{cha} \tag{10}$$

2. Heat production and demand balance

$$\sum_{j=1}^{N_{chp}} H_{jt}^{chp} + \sum_{k=1}^{N_{ho}} H_{kt}^{ho} = H_t^D \tag{11}$$

3. Feasible operating regions of CHP units
There is a dual dependency between generated power and heat of a CHP unit. This dependency is represented using a feasible operation region (FOR) for each CHP. Two types of FORs are distinguished in the existing literature for CHP units. The first type of the FOR can be represented using four extreme points, i.e. A, B, C and D as shown in Fig. 1. This type is mathematically formulated with (12) - (14) [15].

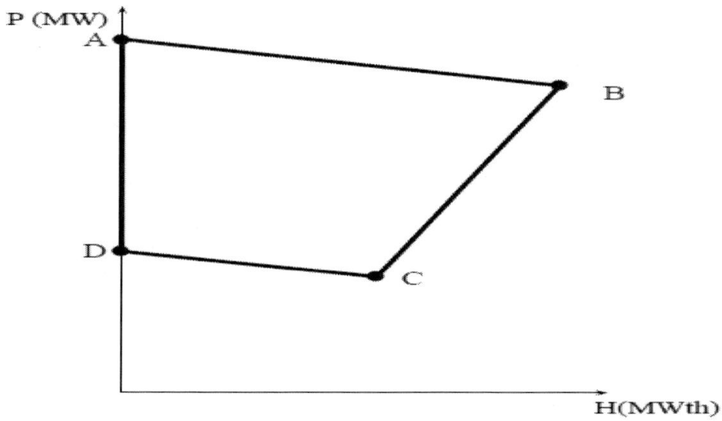

Figure 1. A Typical heat-power feasible region for a CHP unit type I.

$$P_{jt}^{chp} - P_{j,A}^{chp} - \frac{P_{j,A}^{chp} - P_{j,B}^{chp}}{H_{j,A}^{chp} - H_{j,B}^{chp}}(P_{jt}^{chp} - P_{j,A}^{chp}) \leq 0 \quad (12)$$

$$P_{jt}^{chp} - P_{j,B}^{chp} - \frac{P_{j,B}^{chp} - P_{j,C}^{chp}}{H_{j,B}^{chp} - H_{j,C}^{chp}}(P_{jt}^{chp} - P_{j,B}^{chp}) \geq 0 \quad (13)$$

$$P_{jt}^{chp} - P_{j,C}^{chp} - \frac{P_{j,C}^{chp} - P_{j,D}^{chp}}{H_{j,C}^{chp} - H_{j,D}^{chp}}(P_{jt}^{chp} - P_{j,C}^{chp}) \geq 0 \quad (14)$$

The mathematical representation (12) - (14) can be used to model the FOR of a committed unit. However, these equations are not sufficient for the short-term scheduling problem for which the status of units should be determined. Equations (13) and (14) enforces the output power of CHP unit to be more than the values which are determined by FOR borders. Therefore (13) and (14) can not be used for modeling the FOR in the scheduling problem. The equations (13) and (14) can be replaced by (15) and (16) considering that for a decommitted unit ($V_{jt}^{chp} = 0$) the output power is equal to zero. It can be seen that for committed units ($V_{jt}^{chp} = 1$), the new equations is equal to the previous ones. Moreover, the produced heat and power should be set to zero for decommitted units which is enforced by using (17) and (18).

$$P_{jt}^{chp} - P_{j,B}^{chp} - \frac{P_{j,B}^{chp} - P_{j,C}^{chp}}{H_{j,B}^{chp} - H_{j,C}^{chp}}(P_{jt}^{chp} - P_{j,B}^{chp}) \geq -(1 - V_{jt}^{chp}) \times M \tag{15}$$

$$P_{jt}^{chp} - P_{j,C}^{chp} - \frac{P_{j,C}^{chp} - P_{j,D}^{chp}}{H_{j,C}^{chp} - H_{j,D}^{chp}}(P_{jt}^{chp} - P_{j,C}^{chp}) \geq -(1 - V_{jt}^{chp}) \times M \tag{16}$$

$$0 \leq H_{jt}^{chp} \leq H_{j,B}^{chp} \times V_{jt}^{chp} \tag{17}$$

$$0 \leq P_{jt}^{chp} \leq P_{j,A}^{chp} \times V_{jt}^{chp} \tag{18}$$

The second type of FOR for CHP units is depicted in Fig. 2. This type of FOR is represented with six extreme points, i.e. points A-F in Fig. 2. As shown in this figure, the second type of FOR is non-convex. Therefore, this region can not be represented using linear constraints. If the linear constraints are used to represent FOR type II, i.e. (19) - (22), the gray region (FEG region) in Fig. 2 can not be included in the FOR.

$$P_{jt}^{chp} \leq P_{j,A}^{chp} \tag{19}$$

$$P_{jt}^{chp} - P_{j,B}^{chp} - \frac{P_{j,B}^{chp} - P_{j,C}^{chp}}{H_{j,B}^{chp} - H_{j,C}^{chp}}(P_{jt}^{chp} - P_{j,B}^{chp}) \leq 0 \tag{20}$$

$$P_{jt}^{chp} - P_{j,C}^{chp} - \frac{P_{j,C}^{chp} - P_{j,D}^{chp}}{H_{j,C}^{chp} - H_{j,D}^{chp}}(P_{jt}^{chp} - P_{j,C}^{chp}) \geq 0 \tag{21}$$

$$P_{jt}^{chp} - P_{j,E}^{chp} - \frac{P_{j,E}^{chp} - P_{j,F}^{chp}}{H_{j,E}^{chp} - H_{j,F}^{chp}}(P_{jt}^{chp} - P_{j,E}^{chp}) \geq 0 \tag{22}$$

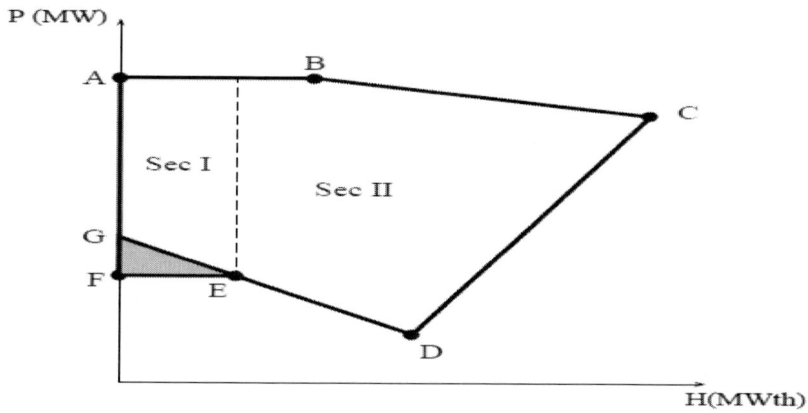

Figure 2. A Typical heat-power feasible region for a CHP unit type II.

$$P_{jt}^{chp} - P_{j,D}^{chp} - \frac{P_{j,D}^{chp} - P_{j,E}^{chp}}{H_{j,D}^{chp} - H_{j,E}^{chp}}(P_{jt}^{chp} - P_{j,D}^{chp}) \geq 0 \tag{23}$$

A possible solution for handling the non-convexity of the second type of FOR is dividing it into convex sub-regions, i.e. sub-regions I and II in Fig. 2 [15]. A binary variable (y_n^{chp}) is required for selecting the active convex region. Hence, constraints (25) - (24) are replaced with (23) - (22) to represent the second type of FOR [15]. Also it should be mentioned that the binary variable are only required for second type of FOR and therefore are shown with y_n^{chp}, where n is an index for the second type of FOR CHPs.

$$(1 - y_n)(P_{jt}^{chp} - P_{j,E}^{chp} - \frac{P_{j,E}^{chp} - P_{j,F}^{chp}}{H_{j,E}^{chp} - H_{j,F}^{chp}}(P_{jt}^{chp} - P_{j,E}^{chp})) \geq 0 \tag{24}$$

$$y_n(P_{jt}^{chp} - P_{j,D}^{chp} - \frac{P_{j,D}^{chp} - P_{j,E}^{chp}}{H_{j,D}^{chp} - H_{j,E}^{chp}}(P_{jt}^{chp} - P_{j,D}^{chp})) \geq 0 \tag{25}$$

Similar to CHP type I, the above mentioned equations can not be used for modeling FOR of CHP in the scheduling problem. On the other hand, the proposed method in [15] has multiplications of binary and continuous variables which makes the problem more complicated. In this chapter a linear equivalent is presented for these equations. Hence, we replace (25) - (24) with (27) - (30). In these equations, binary variables $Y1_{nt}^{chp}$ and $Y2_{nt}^{chp}$ are used instead of y_n^{chp}. $Y1_{nt}^{chp} = 1$ ($Y2_{nt}^{chp} = 1$) means the CHP operates in the first (second) convex section of FOR. Equation (30) enforces the output to be in one of the regions when CHP is ON and in none of them when CHP is OFF. The produced heat and power should be also set to zero for decommited units which is enforced using (31) and (32).

$$P_{jt}^{chp} - P_{j,E}^{chp} - \frac{P_{j,E}^{chp} - P_{j,F}^{chp}}{H_{j,E}^{chp} - H_{j,F}^{chp}}(P_{jt}^{chp} - P_{j,E}^{chp}) \geq -(1 - Y1_{nt}^{chp}) \times M \tag{26}$$

$$P_{jt}^{chp} - P_{j,D}^{chp} - \frac{P_{j,D}^{chp} - P_{j,E}^{chp}}{H_{j,D}^{chp} - H_{j,E}^{chp}}(P_{jt}^{chp} - P_{j,D}^{chp}) \geq -(1 - Y2_{nt}^{chp}) \times M \tag{27}$$

$$H_{jt}^{chp} - H_{j,E}^{chp} \geq -(1 - Y2_{nt}^{chp}) \times M \tag{28}$$

$$H_{jt}^{chp} - H_{j,E}^{chp} \leq (1 - Y1_{nt}^{chp}) \times M \tag{29}$$

$$Y1_{nt}^{chp} + Y2_{nt}^{chp} = V_{nt}^{chp} \tag{30}$$

$$0 \leq H_{jt}^{chp} \leq H_{j,C}^{chp} \times V_{jt}^{chp} \tag{31}$$

$$0 \leq P_{jt}^{chp} \leq P_{j,A}^{chp} \times V_{jt}^{chp} \tag{32}$$

4. System reserve constraint

$$R_t \geq \sum_{i=1}^{N_{po}} (V_{it}^{po} P_i^{po,max} - P_{it}^{po}) + \sum_{j=1}^{N_{chp}} (P_{jt}^{chp,max}(H_{jt}^{chp}) - P_{jt}^{chp}) \tag{33}$$

The system reserve requirement (R_t) is determined based on the system total demand and forecasted intermittent power generation, i.e. wind and PV here. In this chapter the method implemented in [13] is used for determining total reserve requirement. In this method, the total operating reserve is divided into two parts. The first part is for compensating load forecasting uncertainty and is assumed to be fixed percentage of the total system load (e.g., % 5) and the second part is for compensating the wind and photovoltaic power forecasting errors. The second part is assumed to be a percentage of the total dispatched wind and PV power generation (e.g., % 10). Hence, R_t can be expressed mathematically as follows:

$$R_t = k_1 P_t^D + k_2 (\sum_{l=1}^{N_{pv}} P_{lt}^{pv} + \sum_{m=1}^{N_w} P_{mt}^{w}) \tag{34}$$

It should be noted that the maximum power of a CHP unit depends on its produced heat and is determined by the upper bounds of its FOR. These bounds can be implemented using (35) and (36) for CHP type I and (35) and (37) for CHP type II. Equation (35) imposes zero reserve for offline units and is not a constraint for online units. The bounding constraint for CHP type I (II) reserve is (36) ((37)).

$$P_{jt}^{chp,max}(H_{jt}^{chp}) \le P_{j,A}^{chp} \times V_{jt}^{chp} \tag{35}$$

$$P_{jt}^{chp,max}(H_{jt}^{chp}) \le P_{j,A}^{chp} + \frac{P_{j,A}^{chp} - P_{j,B}^{chp}}{H_{j,A}^{chp} - H_{j,B}^{chp}}(H_{jt}^{chp} - H_{j,A}^{chp}) \tag{36}$$

$$P_{jt}^{chp,max}(H_{jt}^{chp}) \le P_{j,B}^{chp} + \frac{P_{j,B}^{chp} - P_{j,C}^{chp}}{H_{j,B}^{chp} - H_{j,C}^{chp}}(H_{jt}^{chp} - H_{j,B}^{chp}) \tag{37}$$

5. Capacity limits of power and heat only units

$$P_i^{po,min} V_{it}^{po} \le P_{it}^{po} \le P_i^{po,max} V_{it}^{po} \tag{38}$$

$$H_k^{ho,min} V_{it}^{ho} \le H_{kt}^{ho} \le H_k^{ho,max} V_{it}^{ho} \tag{39}$$

6. Wind power constraints
The total available wind power of a wind turbine ($P_{kt}^{w,av}$) is a function of the wind speed and turbine characteristics and is expressed as follows:

$$P_{kt}^{w,av} = \begin{cases} 0 & WS_t < V_k^{cutin} \\ P_k^{w,rated} \times \left(\frac{WS_t - V_k^{cutin}}{V_k^{rated} - V_k^{cutin}} \right)^3 & V_k^{cutin} \le WS_t \le V_k^{rated} \\ P_k^{w,rated} & V_k^{rated} \le WS_t \le V_k^{cutout} \\ 0 & WS_t > V_k^{cutout} \end{cases} \tag{40}$$

The generated power of the wind turbine k at hour t is restricted to the available wind power. Also wind and PV power spillage is allowed. The algorithm decides about utilizing renewable generation considering the total cost and the operational constraints. The following constraint enforces this restriction:

$$P_{kt}^{w} \leq P_{kt}^{w,av} \tag{41}$$

7. PV constraints
   The power generated by a PV system is a function of the solar radiation and is expressed as follows [1]:

$$P_{lt}^{pv,av} = A_{l}^{pv} x_t^2 + B_{l}^{pv} x_t + C_{l}^{pv} \tag{42}$$

$$P_{lt}^{pv} \leq P_{lt}^{pv,av} \tag{43}$$

8. Storage constraints
   The energy balance constraint of the storage can be explained as:

$$E_t^s = E_{t-1}^s + \eta_{cha} P_t^{cha} \Delta t - \frac{1}{\eta_{dch}} P_t^{dch} \Delta t \tag{44}$$

The limits of charging power, discharging power and storage capacity are enforced using the following constraints.

$$E^{min} \leq E_t^s \leq E^{max} \tag{45}$$

$$P_{min}^{cha} \leq P_t^{cha} \leq P_{max}^{cha} \tag{46}$$

$$P_{min}^{dch} \leq P_t^{dch} \leq P_{max}^{dch} \tag{47}$$

It should be noted that the battery does not charge and discharge simultaneously because of the additional and unnecessary cost of charge and discharge efficiency deterioration.

3 Illustrative Example

In this section the proposed model is applied to a typical isolated microgrid. At first the structure of the considered microgrid is illustrated and after that the simulation results of optimal operation scheduling is presented.

3.1 Microgrid Structure

The considered microgrid consists of one power-only generation unit (unit I), two CHPs (unit II and III) with different FORs, one boiler (unit IV), PV arrays (aggregated as unit

V), 3 wind units (units VI-VIII) and one NaS battery (unit IX). The cost function of power-only and boiler units are given in (48) and (49). Cost function coefficients and extreme points of CHP units are presented in Table 1 and depicted in Figs. 3 and 4. The forecasted heat and electricity demands are presented in Table 2 for whole planning horizon. The length of operation interval is considered one hour. Expected generation of PV arrays and wind units is based on the forecasted wind speeds and radiations as presented in Table 3. Start up costs of units I - IV and O&M cost of units V-VIII are provided in Table 4. The shutdown costs are assumed to be zero. Characteristics of the battery is presented in Table 5. The initial amount of battery energy level is considered equal to its minimum allowable values. Also the amount of reserve for each hour is considered as 5% of forecasted load and 10% of forecasted renewable energies power, i.e. $k_1 = 5\%$ and $k_2 = 10\%$.

$$C_1(P_1) = 50P_1 \quad 0 \le P_1 \le 15MW \tag{48}$$

$$C_4(H_4) = 23.4H_4 \quad 0 \le H_4 \le 26MWth \tag{49}$$

Table 1. Cost function parameters and FOR extreme points of the CHP units

Unit	a	b	c	d	e	f	feasible region coordinates $[P^c, H^c]$
CHP Unit 1	0.0345	14.5	265	0.03	4.2	0.031	[9.88,0], [8.1,10.48], [21.5,18], [24.7,0]
CHP Unit 2	0.0435	36	125	0.027	0.6	0.011	[4.4,0], [4.4,1.59],[4,7.5],[11.02,13.56], [12.58,3.24],[12.58,0]

Table 2. Forecasted electricity and thermal demands

Time interval (h)	Electricity Demand (MW)	Thermal Demand (MWth)	Time interval (h)	Electricity Demand (MW)	Thermal Demand (MWth)
1	13.51	8.52	13	16.91	13.07
2	12.28	8.97	14	15.37	10.02
3	9.8	6.93	15	19.09	14.48
4	5.23	3.61	16	19.7	15.43
5	7.32	5.14	17	18.16	11.12
6	10.42	10.62	18	19.4	14.38
7	13.51	11.35	19	20.94	12.75
8	15.99	12.3	20	22.18	15.01
9	17.23	10.7	21	22.8	16.34
10	19.09	11.06	22	19.7	16.28
11	19.4	13.7	23	18.78	13.43
12	17.85	10.51	24	17.23	10.47

Table 3. Expected generation of wind units and PV arrays

Time interval (h)	Expected Generation of Wind Unit 1 (MW)	Expected Generation of Wind Unit 2 (MW)	Expected Generation of Wind Unit 3 (MW)	Expected generation of PV arrays (MW)
1	0.44	0.53	0.52	0
2	0.43	0.52	0.53	0
3	0.4	0.49	0.53	0
4	0.36	0.47	0.51	0
5	0.36	0.44	0.5	0.06
6	0.35	0.36	0.5	0.43
7	0.3	0.28	0.46	0.63
8	0.27	0.2	0.37	2.24
9	0.29	0.15	0.3	3.41
10	0.29	0.15	0.28	3.67
11	0.29	0.21	0.29	4.48
12	0.29	0.23	0.28	4.68
13	0.28	0.23	0.27	4.58
14	0.27	0.24	0.25	4.68
15	0.27	0.21	0.25	3.67
16	0.28	0.22	0.27	3.16
17	0.26	0.21	0.28	0.63
18	0.23	0.2	0.3	0.43
19	0.25	0.19	0.31	0.21
20	0.3	0.23	0.33	0.03
21	0.32	0.29	0.4	0.01
22	0.34	0.31	0.45	0
23	0.4	0.29	0.49	0
24	0.44	0.26	0.51	0

Table 4. Start up and O&M costs of units I-VIII

Unit	Start up Cost ($ / start)		O&M cost ($/MWh)
Power only Unit	40	Wind Unit 1	8.2
CHP Unit 1	85	Wind Unit 2	8.1
CHP Unit 2	55	Wind Unit 3	8.5
Heat only Unit	18	PV Unit	4.7

Table 5. Characteristics of NaS battery

Characteristics	Value	Characteristics	Value
E^{Min} (MWh)	1	P^{cha}_{max} (MW)	1
E^{Max} (MWh)	4	P^{cha}_{min} (MW)	0
η_{cha}	0.9	P^{dch}_{max} (MW)	1
η_{dch}	0.9	P^{dch}_{min} (MW)	0

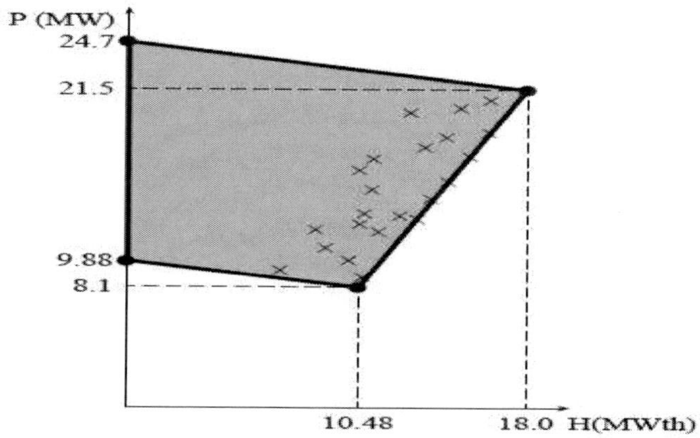

Figure 3. Feasible operation region of CHP unit 1 (unit II).

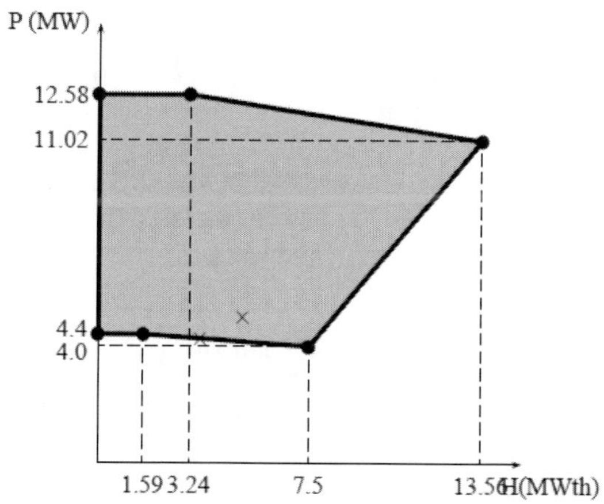

Figure 4. The heat-power feasible regions for a CHP unit 2 (unit III).

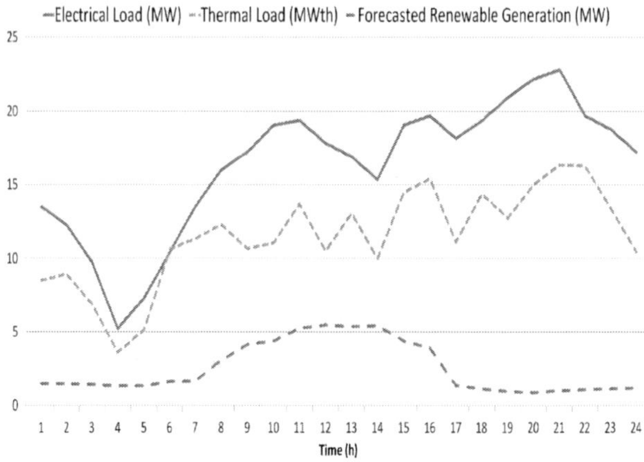

Figure 5. Hourly data of forecasted demand and expected renewable source generations.

3.2 Optimization Results

The problem is modeled in General Algebraic Modeling Language (GAMS) environment [16] and solved using CPLEX solver [17]. The model statistics are shown in Table 6. The problem is solved in less than one second using a typical personal computer. The main reason of this speed is the structure of the model which is formulated as Mixed Integer Quadratic Constrained Programming (MIQCP). The only source of nonlinear terms are multiplication of continuous variables (quadratic functions). The minimum cost of supplying electricity and demand in all planning horizon is $ 16500.47.

Table 6. Model Statistics

# of Equations	1081	# of Binary Variables	336
# of Variables	721	# of Continuous Variables	385
Duality Gap (%)	0.00%	Solution Time (sec)	0.944

The power and heat output of units are presented in Tables 7 and 8, respectively. As shown in these tables, both power-only and heat only units are not used to produce power and heat. The electricity demand is supplied with the renewable resources and CHP units for all time periods and with battery in some hours. Also the heat demand is provided using CHP units. The operation point of CHP units in different hours of planning horizon are shown on FOR regions in Fig. 3 and Fig. 4 with × sign. The first CHP unit is used in hours 1 to 3 and 6 to 24. The second CHP unit is used in hours 4 and 5. For hours 4 and 5, the demand minus forecasted renewable energy is greater than the maximum power of first CHP unit, therefore the first CHP unit is turned off in hour 4 and the second CHP unit is started up to supply electricity and heat demands.

Table 7. Hourly power output of units and system reserve

Time interval (h)	CHP_1 Power (MW)	CHP_2 Power (MW)	Renewable Generation (MW)	Reserve (MW)
1	12.02	0	1.49	0.82
2	10.8	0	1.48	0.76
3	9.24	0	1.42	0.63
4	0	4.26	1.34	0.4
5	0	4.96	1.36	0.5
6	8.78	0	1.64	0.68
7	11.84	0	1.67	0.84
8	12.91	0	3.08	1.11
9	13.08	0	4.15	1.28
10	14.7	0	4.39	1.39
11	14.13	0	5.27	1.5
12	12.37	0	5.48	1.44
13	12.72	0	5.19	1.36
14	9.93	0	5.44	1.31
15	15.23	0	4.4	1.39
16	16.92	0	3.78	1.36
17	16.78	0	1.38	1.05
18	18.24	0	1.16	1.09
19	19.98	0	0.96	1.14
20	20.29	0	0.89	1.2
21	20.78	0	1.02	1.24
22	18.54	0	1.1	1.09
23	17.6	0	1.18	1.06
24	16.02	0	1.21	0.98

The electricity and heat balance for all hours are shown in Fig. 6 and Fig. 7, respectively. In these figures, generations are shown in the upper part (as positive values) and demands in lower parts (as negative values).

Battery energy levels and charging/discharging rates during the operation hours are shown in Fig. 8 and Fig. 9, respectively. In hours 13 and 16 some renewable generation spillage is occurred. The spillage is 0.17 MW for hour 13 and 0.15 MW for hour 16. As can be seen in Fig. 9, in hours 13 and 16 the storage charging rate is equal to its maximum. Also for the first CHP unit, the generated power is equal to its minimum value considering the generated heat (see Fig. 3).

Table 8. Hourly heat output of CHPs

Time interval (h)	CHP_1 Heat (Mwth)	CHP_2 Heat (Mwth)	Time interval (h)	CHP_1 Heat (Mwth)	CHP_2 Heat (Mwth)
1	8.52	0	1	13.07	0
2	8.97	0	2	10.02	0
3	6.93	0	3	14.48	0
4	0	3.61	4	15.43	0
5	0	5.14	5	11.12	0
6	10.62	0	6	14.38	0
7	11.35	0	7	12.75	0
8	12.3	0	8	15.01	0
9	10.7	0	9	16.34	0
10	11.06	0	10	16.28	0
11	13.7	0	11	13.43	0
12	10.51	0	12	10.47	0

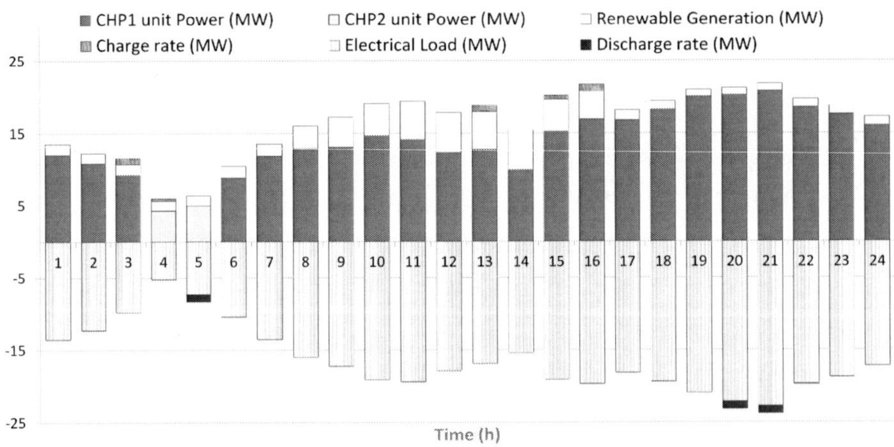

Figure 6. Hourly power generation and demand balance.

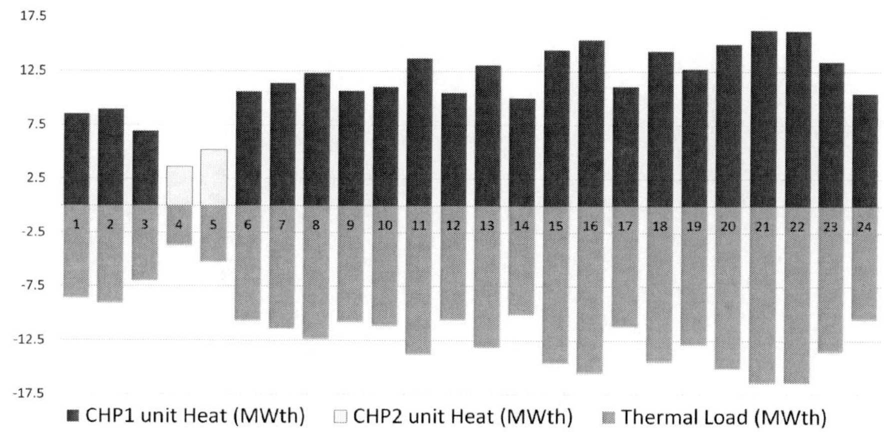

Figure 7. Hourly heat generation and demand balance.

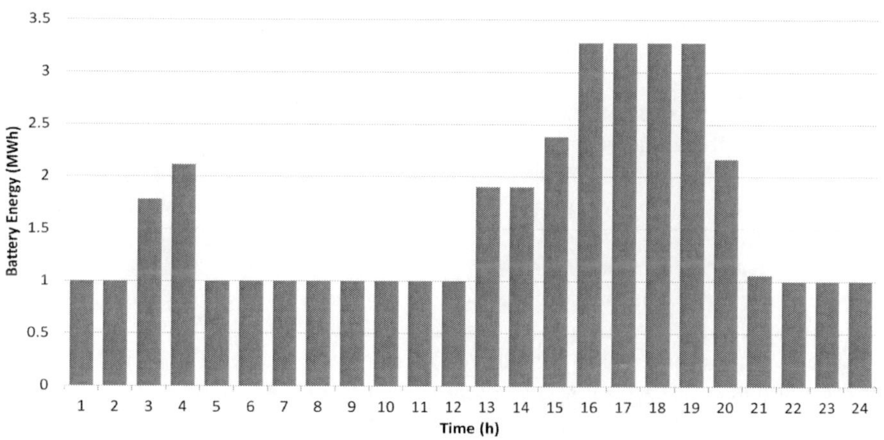

Figure 8. Battery energy in operation horizon.

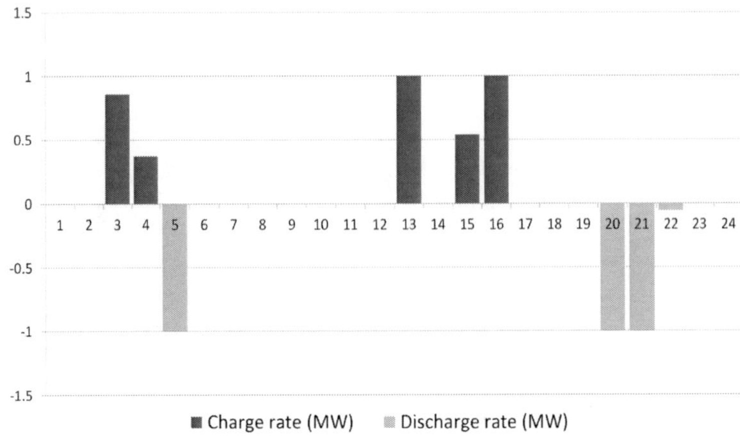

Figure 9. Battery charging and discharging rates.

Conclusion

In this chapter optimal short-term operation scheduling of an isolated microgrid with CHP and renewable generation is studied. The considered microgrid has thermal and power consumptions. It is assumed that the thermal load can be served using CHP units and/or heat-only (boiler) units. The power demand can be supplied by CHP units, power-only units, wind turbine, PV and battery storage. The effect of the uncertain wind and solar power productions are considered in determining the system spinning reserve requirement. The short-term scheduling of microgrid is formulated as a mixed integer quadratically constrained program (MIQCP) and has been solved using GAMS software. A linear and mixed integer equivalents are proposed for representation of convex and non-convex feasible operation regions of CHPs. An illustrative example is developed to demonstrate the efficiency of the proposed method. As a future work, the model in this chapter can be extended by including the uncertainties with a stochastic programming technique.

References

[1] M. Mohammadi, S.H. Hosseinian, and G.B. Gharehpetian. Optimization of hybrid solar energy sources/wind turbine systems integrated to utility grids as micro-

grid (mg) under pool/bilateral/hybrid electricity market using pso. *Solar Energy*, 86:112–125, 2012.

[2] Prasenjit Basak, S. Chowdhury, S. Halder nee Dey, and S. P. Chowdhury. A literature review on integration of distributed energy resources in the perspective of control, protection and stability of microgrid. *Renewable and Sustainable Energy Reviews*, 16(8):5545–5556, 2012.

[3] Xian Liu. Optimization of a combined heat and power system with wind turbines. *International Journal of Electrical Power and Energy Systems*, 43(1):1421–1426, 2012.

[4] Aiying Rong, Henri Hakonen, and Risto Lahdelma. A dynamic regrouping based sequential dynamic programming algorithm for unit commitment of combined heat and power systems. *Energy Conversion and Management*, 50(4):1108–1115, 2009.

[5] Aiying Rong and Risto Lahdelma. An effective heuristic for combined heat-and-power production planning with power ramp constraints. *Applied Energy*, 84(3):307–325, 2007.

[6] Fabricio Salgado and Pedro Pedrero. Short-term operation planning on cogeneration systems: A survey. *Electric Power Systems Research*, 78(5):835–848, 2008.

[7] A. Kahrobaeian and Y. A. R. I. Mohamed. Interactive distributed generation interface for flexible micro-grid operation in smart distribution systems. *IEEE Transactions on Sustainable Energy*, 3(2):295–305, 2012.

[8] Behnam Mohammadi-Ivatloo, Mohammad Moradi-Dalvand, and Abbas Rabiee. Combined heat and power economic dispatch problem solution using particle swarm optimization with time varying acceleration coefficients. *Electric Power Systems Research*, 95(0):9 – 18, 2013.

[9] Andreas Christidis, Christoph Koch, Lothar Pottel, and George Tsatsaronis. The contribution of heat storage to the profitable operation of combined heat and power plants in liberalized electricity markets. *Energy*, 41(1):75–82, 2012.

[10] Aiying Rong, Risto Lahdelma, and Peter B. Luh. Lagrangian relaxation based algorithm for trigeneration planning with storages. *European Journal of Operational Research*, 188(1):240–257, 2008.

[11] M. Abbaspour, M. Satkin, B. Mohammadi-Ivatloo, F. Hoseinzadeh Lotfi, and Y. Noorollahi. Optimal operation scheduling of wind power integrated with compressed air energy storage (caes). *Renewable Energy*, 51:53–59, 2013.

[12] M. A. Fotouhi Ghazvini, Hugo Morais, and Zita Vale. Coordination between mid-term maintenance outage decisions and short-term security-constrained scheduling in smart distribution systems. *Applied Energy*, 96(0):281–291, 2012.

[13] H. Siahkali and M. Vakilian. Stochastic unit commitment of wind farms integrated in power system. *Electric Power Systems Research*, 80(9):1006 – 1017, 2010.

[14] Hugo Morais, Peter Kadar, Pedro Faria, Zita A. Vale, and H.M. Khodr. Optimal scheduling of a renewable micro-grid in an isolated load area using mixed-integer linear programming. *Renewable Energy*, 32:151–V156, 2010.

[15] Z.W. Geem and Y.H. Cho. Handling non-convex heat-power feasible region in combined heat and power economic dispatch. *International Journal of Electrical Power & Energy Systems*, 2011.

[16] A. Brooke, D. Kendrick, A. Meeraus, and R. Raman. *GAMS: A user's guide*. GAMS Development Corporation, 1998.

[17] A. Brooke, D. Kendrick, A. Meeraus, and R. Raman. *GAMS/Cplex 7.0 User Notes*. GAMS Development Corp., 2000.

In: Smart Microgrids
Editor: Hussein M. Khodr

ISBN:978-62808-975-2

Chapter 3

DEMAND RESPONSE UNDER SMART GRID PARADIGM

V. S. K. Murthy Balijepalli, Vedanta Pradhan and S. A. Khaparde*
Department of Electrical Engineering, Indian Institute of Technology, Bombay, India

1 Nomenclature

DR	Demand Response
LSE	Load Serving Entity
CSP	Curtailment Service Provider
ISO	Independent System Operator
RTO	Regional Transmission Operator
TSO	Transmission System Operator
IL	Interruptible Load
$PHEV$	Plug-in Hybrid Electric Vehicles
$HVAC$	Interruptible Load
EE	Energy Efficiency
DSM	Demand Side Management
DMS	Distribution Management System
DRP	Demand Response Provider
DSO	Distribution System Operator
MO	Market Operator
DG	Distributed Generation

*e-mail(s): {vsk, vedanta}@ee.iitb.ac.in

DER	Distributed Energy Resources
LMP	Locational Marginal Pricing
t	Index for time block
T	Total number of time blocks in a day
A	Set of all appliances
λ	Price of electricity Rs/kWh
λ_o	Nominal price of electricity Rs/kWh
ρ	Total consumption from LSE kWh
ρ_o	Nominal overall demand when price is nominal kW
ρ_a	Consumption of appliance a kWh
E_a	Average daily consumption of appliance 'a' kWh
α_a	First hour of operation time-band for appliance 'a'
β_a	Last hour of operation time-band for appliance 'a'
PJM	PJM Interconnection LLC
ρ_a^{min}	Minimum consumption level of appliance 'a' in its operational time band kWh
ρ_a^{max}	Maximum consumption level of appliance 'a' in its operational time band kWh
B_o	Nominal benefit when price is nominal Rs
$B(\rho_t)$	Customer benefit during time block t from the use of ρ_t of electric energy Rs
E	Average self price-elasticity of overall demand
$LBNL$	Lawrence Berkeley National Laboratory
$SDG\&E$	San Diego Gas & Electric
$DRRC$	Demand Response Research Center
$UCAIug$	UCA International Users Group
$OASIS$	Group Promoting Open Standards for the Information Society
CPP	Critical Peak Pricing
$NAESB$	North American Energy Standards Board
$NYISO$	New York Independent System Operator
$CAISO$	California Independent System Operator

Demand Response (DR) attains a crucial role in the context of the operations and economy of the smart grids. Traditionally, grid security management has been done mostly by the utility side resources of real and reactive power with the demand side taking limited responsibility in them. With the ever increasing base and peak electricity consumption scenarios, availability of these generation side resources naturally tend towards a bottle neck.

At this juncture, the role of active demand side participation in the operations of the grid stability and security becomes inevitable. DR is a collective term for any action taken from the demand side in a proactive or reactive way to cater to the needs of the grid. Although specifically, voluntary load management, load control, peak shaving and peak shifting have been the flag bearing applications of DR, the scope of research for other DR programs and related business models is ample. Smart grids transformed the existing DR programs to fully automated schemes with the inception of open data standards. This chapter provides a bird's eye view of DR in all of its aspects covering the concept, the enabling technology, the mathematics involved, the global standards for implementation, challenges and the future research. Authors have tried to cover all these aspects to the extent of detail, possible for this chapter. Interested readers can follow up for further details by the references provided at the end of the chapter.

2 Background

Technology is recognized as a vital ingredient for transforming the present day power grid to the needs of 21st century consumers. Demand response has been operated traditionally with limited objectives and lack in consumer participation. To overcome the present difficulties, activities on Smart Grids has been reported in some distribution pockets mostly by private players. Many utilities in power sector have now Smart Grid activity with high priority in their research and development agenda. Smart grids initiatives are being taken up throughout the world to improve electricity grid performance and increase the role of end-consumer in the power ecosystem. This concept integrates energy infrastructure, processes, devices, information and markets into a coordinated and collaborative process that allows energy to be generated, distributed and consumed more effectively and efficiently. Though the deployment strategy for Smart Grid encapsulates all stages of the electric power life cycle (generation, transmission, distribution, metering, customer), it should initially focus on sustainable options for customers at the distribution, meter and customer phases. Hence, a Smart Grid should enable devices at all levels within the grid (from utility to customer) to independently sense, anticipate and respond to real-time conditions by accessing, sharing and acting on real-time information. Simply stated, Smart Grid is the convergence of information and operational technology applied to the electric grid, allowing sustainable options to customers and improved security, reliability and efficiency to utilities. A modernized grid would create a digital energy system that will:

- Improve grid reliability, security and efficiency
- Respond to local and system-wide inputs and have much more information about broader system problems
- Facilitate dynamic optimization of grid operations and resources
- Meter and communicate grid status/operation
- Provide cyber security
- Increase the involvement of end-consumers in the energy supply chain
- Provide to consumers with timely information and control options
- Advance demand response/management
- Integrate "smart" appliances and consumer devices like hybrid cars, electric vehicles, etc.
- Provide a flexible platform for integration of energy storage and renewables
- Enhance deployment and integration of distributed resources and generation including mobile - electric vehicle/stationary storage
- Further distribution automation

It is identified that one of the ways to succeed in implementing a Smart Grid that can enable a wide range of intelligent applications well into the future is the usage of "standards based approach" and the focus on dual interoperability of technologies. In this context, this chapter will introduce the reader about the concept of DR, literature on DR, Mathematical models to the standardization activities for promoting interoperable DR.

Organization of the Chapter

With this background about smart grids we introduce the topic of Demand Response (DR) straightforwardly. This chapter will deal with the basics of DR; what it is and how it works. A broad classification of existing DR programs is provided to uncover the fundamentals behind such different programs. The multi-domain benefits of ushering in DR in the existing electricity grids are listed out to bring out its actual importance. As pointed out earlier, a smart grid entails a plethora of technological advancements, various enabling technologies which support realization of DR are discussed. Further, to provide a theoretic flavor to the topic, the mathematical perspective of DR is presented. A few mathematical models of DR are explained in detail. Simulation results with practical data are discussed to bring out the value of DR in terms of actual numbers and figures. The tail end of the chapter covers a few DR pilots and implementations across the globe, challenges in practical implementation, future of DR and research opportunities. The chapter closes on a very important aspect of DR i.e. standardization of DR technology. A power system network scale DR entails information storage and exchange among different players. In order to enable interoperability among these players standardization of DR information technology is a must.

3 Introduction: What It Is and How It Works?

3.1 Popular Definitions

Demand Response is going to become a part of the system operations in the smart grid driven restructured power system around the world in the near future. DR implementations are more active at the retail level than the wholesale level. To enhance competition at the retail level, separate entities called retailers have also come into the scenario. The increased retail level competition is associated with a variety of problems which can be categorized as market based and network based problems [30]. The former problems occur when the generators or the retailers face financial risks caused by spot price volatility in the wholesale electricity market. The latter problems occur when TSO and DSOs have to maintain reliable power supply during times of peak demand or low operating reserves or when constrained networks are operating at their limits.

Traditionally, problems of the latter type have been handled single sidedly, by the generating utilities who have to either ensure a security margin of generation to be always available to be dispatched when asked to do so by the ISO or in the opposite case have to reduce generation to bring the network from a constrained state to the normal state. A resource which is left unused is the demand side resource which can also be helpful in such situations.

Demand Side Management was introduced by Electric Power Research Institute (EPRI) in the 1980s. DSM is a global term that includes a variety of activities such as: load management, energy efficiency, energy saving, etc. The problems mentioned above can be categorized as short term problems whereas problems such as environmental effects of burning coal to produce electricity can be categorized as long term problems. DSM schemes like energy efficiency and energy saving schemes are potential inhibitors of such problems whereas the short term problems can be tackled by efficient load management programs which are collectively referred to as Demand Response (DR). DR put in simple terms is a collective term for any action taken from the demand side in a proactive or reactive way to cater to the needs of the grid. Standard definitions of the same are as follow:

According to Federal Energy Regulatory Commission, Demand Response (DR) is defined as:
"Changes in electric usage by end-use customers from their normal consumption patterns in response to changes in the price of electricity over time, or to incentive payments designed to induce lower electricity use at times of high wholesale market prices or when system reliability is jeopardized." The end-use customers can also be the customers who are participating in the wholesale market operations.

According to the Sixth Northwest Conservation and Electric Power Plan [105], DR is defined as,
"A voluntary and temporary change in consumers' use of electricity when the power system is stressed. The change in use is usually a reduction, although there are situations in which an increase in use would relieve stress on the power system and would qualify as DR."

3.2 Generic Classification of DR

DR programs can be broadly categorized along two dimensions [97] as shown in Figure 1. The first dimension characterizes how and when LSEs call on program participants to shed load. The second dimension shows the method by which utilities motivate their customers to participate. Participants can be called upon for either emergency/reliability conditions or for economic purposes. Emergency programs offer customers payments for reducing their electricity demand during system contingencies such as generator or line failures. Economic programs offer customers incentives to reduce loads during non-emergency periods when LSE cost of service exceeds some specified limit.

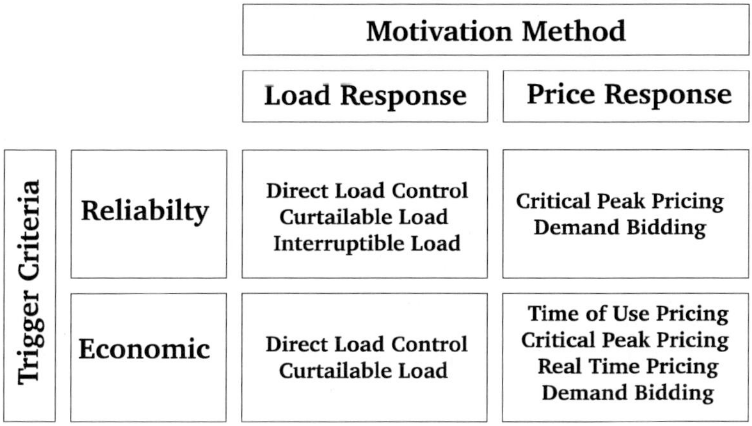

Figure 1. A conceptual classification of demand response programs.

The second dimension concerns the methods by which LSEs motivate their customers to shed load. LSEs may use suitable communication signals to notify participants of reliability or economic events that merit load response in exchange for direct utility payment. Alternatively, they may use pricing signals to more accurately reflect the cost of providing power as a function of time. In price response, LSEs structure their rates to go up during hours of either peak demand or supply constraint. Rates are kept lower at all other times. In real-time pricing, utilities expose customers directly to the full and fluctuating range of wholesale power market prices.

A physical classification of DR programs can be made depending on how the changes in electric usage are implemented. It is shown in Figure 2.

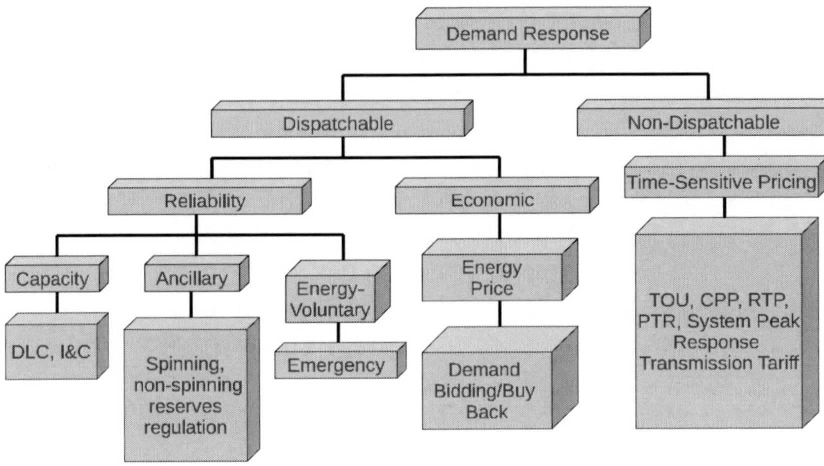

Figure 2. Physical classification of DR schemes.

3.3 Benefits of Demand Response

Benefits of DR are multi-faceted. Successful DR implementation can usher in operational and efficiency enhancements in many sectors of the power system. The following is a list of the notable benefits of DR.

- DR can reduce system peak load in the long term and therefore postpone the need for building new power plants, leading to considerable environmental impacts.

- DR can improve reliability of the transmission network which results from reducing the probability of forced outages. By dispatching DR services at critical times (e.g. when a generator or a transmission line is unexpectedly lost), the TSO can help to return system reserves to pre-contingency levels. Moreover DR can be dispatched in less than 5 minutes and can be better substitutes to peaking power plant which take up to 30 minutes to ramp up to full capacity.

- Relieving the voltage constrained power transfer problem and congestion in the distribution substations, simplifying outage management and improving the quality of supply are the benefits that DR can provide at the distribution network level. With lower power outages, the loss of global monetary value due to non-supplied load can be reduced.

- Retailers buy electricity from the wholesale market and sell it to their consumers at a flat rate. So they are exposed to the financial risks involved with the spot price volatility in the real-time horizons. In order to cover most of these risks they can ask their consumers to reduce their consumption during the times when spot prices are most volatile and reach their peaks, provided the customers can receive financial reward for such decrease in consumption.

- The short term impacts of DR on electricity markets leads to financial benefits of both the utility and the consumers.

- Deployment of new technologies like, distributed generation (solar, small wind, geothermal) and storage (stand-alone, PHEV) also motivate the inclusion of DR as a key component for the smart grid. For example during times of high wind speeds, the generation is quite in excess. A curtailment in wind generation proves to be inefficient for the wind farm, making its payback period quite high. So DR can be used to increase the demand during such periods.

3.4 DR Transformations Under Smart Grids

DR Transformations Under Smart Grids compared to the conventional DR programs is summarized in Table 1.

3.5 Smart Grid Enabling Technologies for DR

The following are globally acknowledged as the enabling technologies for DR-

3.5.1 Smart Meter:

A smart meter is typically a solid state advanced meter which is capable of interval metering. Apart from accurate interval metering, smart meters also

Table 1. Demand Response Transformations

	Conventional DR	**Smart Grid DR**
Participation	Targeted, Limited to Large commercial/Industrial & Residential	All Consumers
Who Controls	Utility	Customer
What is Controlled	• Interruptible rates • Res. HVAC, Water Heating	All Loads Available
Control Equipment	• Utility Provided • Few Suppliers	• Customer Provided • Many Market Suppliers
Incentives	• Fixed Participation Payments • Baseline Metrics	• Retail Dynamic Prices • Reservation Payments • Pay-for Performance
DR Products	Generally Limited to Reliability	Capacity, Energy, Ancillary Services Markets; Congestion Management
DR, EE, Renewable Integration	No	Yes

entail a host of other applications. These are listed as below:

- Measure and record monthly kilowatt hour usage and kilowatt demand
- Measure and record interval (daily, houlry, sub-hourly) usage data
- Remote provision of these data to customers, suppliers and other designated parties
- Outage and tamper notification
- Support Time of Use and other real-time pricing schemes
- Support remote disconnection service
- Be IP addressible and integrable with a Home Area Network (HAN)
- Communicable over local area network (LAN) with local data concentrators via suitable modules

A HAN comprises of a small range communication network, to facilitate communication of signals within the several enabling devices in the customer premise. Zigbee, wifi, Power Line Communication (PLC) using electrical wiring of the premise, provide suitable options for deployment of a HAN.

3.5.2 In-Home Display (IHD) Device

IHDs enable customers to track their energy usage in chart or graph form. Additionally, screens are customizable to provide any type of energy awareness information like carbon footprint data, total energy consumption, and comparative analysis with typical energy use. Also, utility side real-time messages regarding billing, energy prices, etc. can be received. The IHD can control thermostats and adjust their setpoints. It can also control Load Control Modules for appliances like pool pumps, water heaters, lighting, etc.

IHD units can acquire consumption information from different loads through a sensor with built-in RF and/or PLC. However, information sharing with the smart meter through a HAN is more effective. IHD units come with interactive user interface with touch screen. They support displays, connectivity via universal standard bus (USB) and radio frequency (RF), and consume low power.

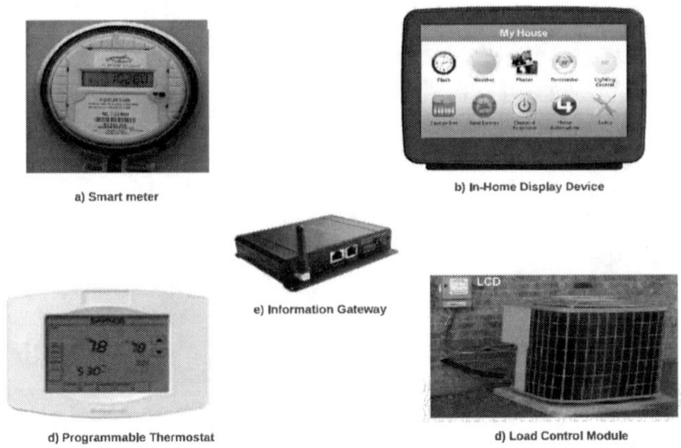

Figure 3. Smart Grid enablers of Demand Response.

3.5.3 Programmable and Controllable Thermostats (PCTs)

Heating, Ventilation and Air Conditioning (HVAC) loads constitute a major load in any consumer premise. Controllable thermostats provide the flexibility to control the thermostat settings of the HVAC loads in a manual or automatic manner. A PCT on the other hand, allows programming of the thermostat control according to dome criteria. The driving factor for this may be the time varying price of electricity over the day, the availability of inhabitants in a premise or the natural requirements of the consumer according the ambient temperatures. A PCT may be interfaced with a Load Control Module for control of thermostat settings, through the HAN or may have an independent communication module for direct communication with the smart meter.

3.5.4 Load Control Module

Load Control Modules fit in with appliances or other load devices in order to control the load demand by them. A typical load control module comes with the communication module, the algorithm module and the associated hardware component. Once interfaced with an appliance or device, this module helps coordinate consumption according to DR signals. LCDs are indispensable in

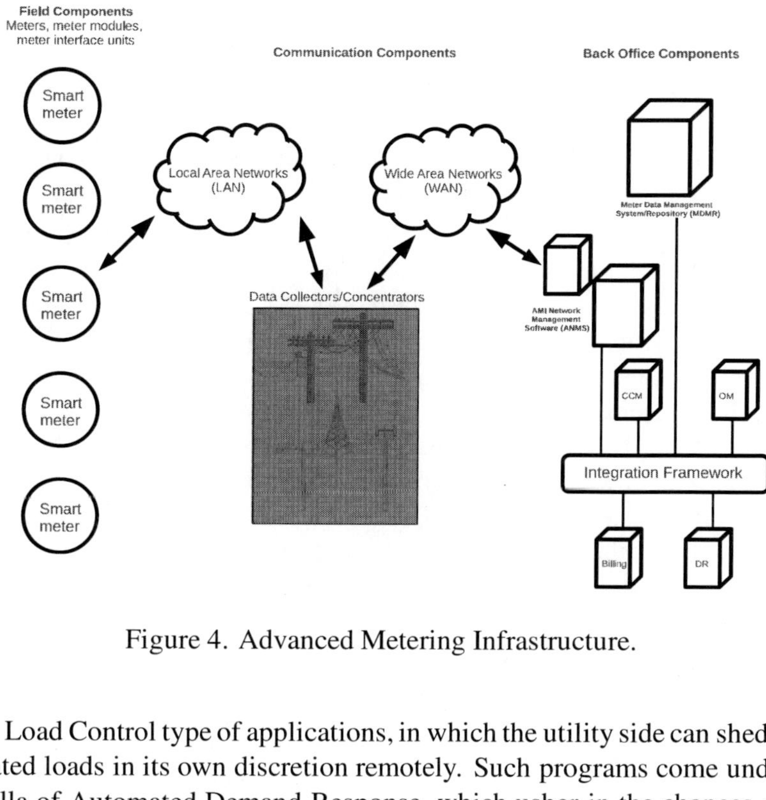

Figure 4. Advanced Metering Infrastructure.

Direct Load Control type of applications, in which the utility side can shed some dedicated loads in its own discretion remotely. Such programs come under the umbrella of Automated Demand Response, which usher in the chances of DR participation in electricty markets.

3.5.5 Advanced Metering Infrastructure (AMI)

The AMI system comprises of all three components of a smart consumer-supplier interaction system. It consists of the smart field meters, meter modules and interfacing units at the consumer side and the meter data management system (MDM), network management software (ANMS) and other back office processes at the utility end. The meter data is communicated from the consumer premises to the local data collectors/concentrators via suitable Local Area Networks (LAN). Suitable Wide Area Networks (WAN) are then used to communicate these data from these data collectors to the utility end. LAN architectures like wireless broadband, wireless mesh networks, licensed or unlicensed wireless (private/public) and Power Line Communication (PLC) and Broadband over Power Line (BPL) are possible options. The wide area network (WAN) portion [96] of the system can use telephone, cellular or other public sources such as satellite, licensed frequencies, optical fiber or private radio networks or microwave to backhaul the data from the collector to the data center.

The MDM interfaces with the AMNS and use the stored meter reading and other AMI data for applications like Customer Information System (CIS), Billing, Geographical Information System (GIS), distribution automation and other operations applications. MDM may be used to bill time of use, critical peak pricing, interruptible service and other complex rates. The combination of [96] the company's CIS system and the MDM application is used to handle billing of all time-differentiated rate structures including the handling of switching functions required to support retail competition and the associated reconcilement process. The implementation of an MDM system is generally synchronized to take place at, or near the beginning of, an AMI system implementation in order to establish early interfaces with the company's CIS and outage management functions and to facilitate the management of large volumes of customer meter change data that will eventually result. The MDM can also provide information concerning the performance of the AMI system and the health of the newly-installed meters by performing validation, editing, and estimating of the data on a daily basis after it has been received by the AMI system.

3.6 Standardization for DR

Grid capacity constraints are growing quickly and will only intensify as more renewable energy sources are brought on line. Intelligent DR automation is a key component to maintaining grid stability, and also the DR schemes should be interoperable with the energy management systems available at the utility-side. For which, the standardization of DR is very essential. OpenADR, OASIS and IEC PC118 are working towards fulfilling these expectation by coordinating with the customers, utilities and the industry. More details about these standardization bodies is presented in the later sections.

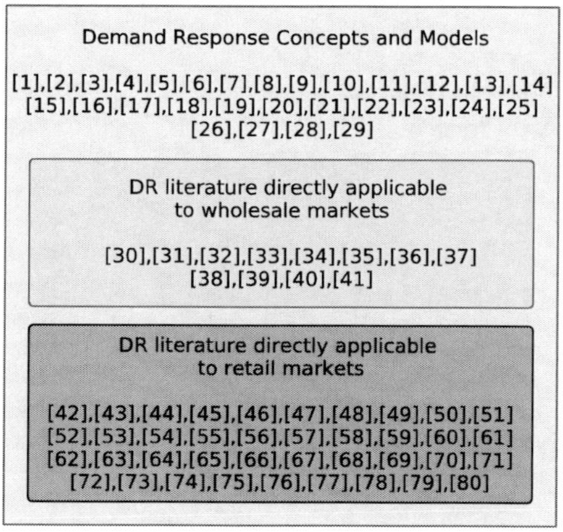

Figure 5. Publications classification.

4 Literature Review on Demand Response

The available literature on DR is classified as: Literature on DR concepts and models, literature on DR frameworks directly applicable to wholesale markets, literature on DR frameworks directly applicable to wholesale markets. Figure 5 gives an overview of the papers classified on the basis of aforesaid notion. The number of major journal and conference publications from year 2000 to July 2011 is shown in the Figure 6. It is estimated that a total of 250 publications on DR can be available by the end of year 2011.

A category-wise discussion of literature follows.

4.1 Literature on DR Concepts and Models

The literature in this category mainly talks about the concepts which guide incorporation of DR into the system operation at all operational time-scales. Various models for scheduling or dispatching DR which can be extended to either wholesale or retail markets are presented. Some papers also present measure of the effects of inclusion of DR into system operation. Reference [1] introduced the economics theory of the electricity industry. The old "vertical integration

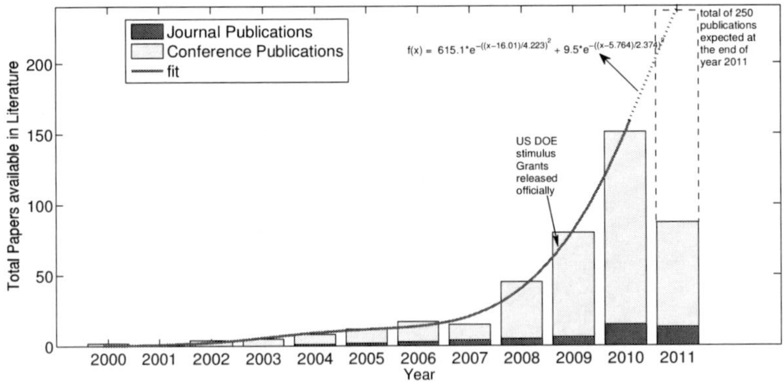

Figure 6. Journal and conference publications chronology.

model" (where central planners set the generation capacity levels, which is an inefficient solution) and the new market-based ones are presented. The problems related to transitioning to the new model are analyzed. The market-based model relies on energy prices to achieve reliability (expressed as outage probability) and short-run efficiency in production and consumption. To work properly, effective DR mechanisms are needed: consumers must be charged on an hourly basis and vary consumption with a certain level of responsiveness according to marginal prices (active demand side). If DR is inadequate, socialized reliability solutions (such as installed reserve margins) become necessary. In reference [2], a review of DSM techniques and a list of future challenges are presented. Regarding dynamic pricing, the authors states that "if prices are not sufficiently different, it will be difficult to justify investments on DSM". More details about the real-time electricity pricing with the possible avenues for their implementation are discussed in [3,4]. Authors in reference [5] have carried out mathematical analysis of the side-effects of DR apart from the desirable result of reducing the local market price of electricity on a simple 3-bus power system. The importance of simultaneous plant production and demand reduction scheduling for the establishment of a fully competitive market is emphasized in [6]. A shift of thought from demand response being negative generation to redistributable demand is proposed. The optimization problem is formulated based upon the cost of producing electricity which consists of the sum of cost of generating it and the cost of demand reduction. The constraints involved are limits on generator capacities and demand reduction amount, the total available energy reduction from a particular Demand Side Bidder (DSB), and the supply and demand balance for every scheduling period. But the paper fails to address the network operating constraints. In reference [7], authors proposed a method for optimally selecting the locations of DR alongwith their capacities to achieve a number of objectives like maximizing available transfer capacity, minimizing expected energy not supplied, minimizing active power loss and minimizing total DR program capacity. In case of such multiple objectives which are conflicting in nature, the NSGA II algorithm is used to obtain pareto optimal solution, for a given domain of possible solutions.

In reference [8] an attempt is made to evaluate the effects of implementing demand response on system variables like total cost of energy and market clearing price by formulating a unit commitment problem. Reference [9] proposed a method of load control based on the price elasticity of consumers. The load control event is triggered by the need of the supplier to reduce the load. The optimization problem is minimization of the cost to the consumers subject to constraints like maximum load reduction for a particular customer, maximum increase in electricity price, and the load and generation balance. A simulation model involving the model of the grid incorporating the aggregate generator inertia, governor action, and load-frequency dependence, model of large number of refrigerators (as frequency responsive loads) and the model of a dynamic demand controller (DDC) has been developed in [10] for investigating whether frequency stability could be provided by incorporating DDC into consumer appliances.

Authors in [11] examined the use of real-time electricity pricing in the domestic sector in supporting the penetration of wind generation. The basic idea being that during periods of low wind generation i.e. when only thermal generation is available and the forecasted load is close to the supply capacity, then the real-time price signal will be high as per the pricing model and thus bring in DR. Authors in [12] discussed the viability of a reliability based rather than a price-based DR program as the Korean electricity market is an incomplete market as price signals are not informed to the customers. Moreover the demand bidding market is settled after the generation scheduling plan is done. The incentive to the participants of the proposed reliability based DR programs comprises of an energy reward, a reserve reward and a capacity reward which must be set so as to fully compensate the participants of their cost of load reductions.

Reference [13] proposed the modeling and integration of the Distribution system topology into the DR scheduling process. This allows calculation of DR available at the nodes of the transmission thus helping in better monitoring and verifying of requested DR. This paper also proposes the addition of a Distribution System Operator to analyze the effects of DR events on the distribution grid. Reference [14] proposed the implementation of DR from the DMS level instead from an aggregator or DRP or substation level in order to achieve the additional benefits of DR like efficient congestion relief in the distribution network, satisfying the voltage constraints of distribution system buses while achieving load reduction which may be violated otherwise. Reference [15] explored the various issues related to the two forms of demand response namely, reliability based and price based demand response. It claims that neither of them can individually maximize the benefits out of demand response.

DR can be considered as a virtual resource which can be exchanged between two groups of participants, retailers, DSO, TSO and MO on one hand and the aggregators and consumers on the other. In this respect two business models are proposed in [16] namely, bilateral and pool based. Authors in reference [17] investigated the performance of centralized and decentralized approaches for DR on a modified IEEE 34 bus system with detailed household load models including dynamic models for air-conditioners, water heaters as well as an aggregated load profile for other loads. The effects of these lower level distribution loads on the transmission level operations can be easily derived after reading the paper. A small islanded microgird with 3.125 MW, 2.4 KV diesel generator, equipped with speed governor and exciter, as a DG alongwith fixed active and dynamic loads are modeled in MATLAB/simulink environment in [18] for simulating frequency and voltage regulation events and observing the recovery due to the response from flexible loads which are centrally controlled by a controller based on Adaptive Hill Climbing technology (AHC). Simulation results show that such control actions improve the transient part of the frequency profile under sudden load disturbances.

A primary step in DR implementation is the detailed knowledge of customer potential through customer aggregation and the characterization of the demand clusters. For example customers whose demand follow the day ahead or the real-time prices are the ones who are more suitable for DR programs selection from the view point of both the customers and suppliers. So, the effective contribution of these classes of customers needs detailed knowledge of customer potential through customer classification and their characterization. Clustering provides the typical load pattern of each customer class which can then be used for applications like choosing suitable DSM programs, tariff structures, etc. and demand characterization or modeling provides an effective tool to estimate the potential demand reduction, loss of service cost and the impact of DR programs on demand. In literature there are a number of clustering techniques like K-means, self-organizing maps, modified follow the leader, etc. [19] presented a modification on the K-means and the fuzzy average K-means to make it more suitable to DR applications. Clustering Dispersion Indicator is proposed as a measure of adequacy of the methods on the basis of which these methods are compared after implementing them on the data of 316 load curves of non-residential customers of Tehran Distribution network. In reference [20], the use of the ISODATA algorithm for customer classification is proposed which also includes temperature dependency correction and outliers filtering. The applicability of the algorithm on the practical data is questionable. Reference [21] proposed the integration of tools like self-organizing maps and physically based load modeling to achieve the above mentioned requirements. [22] presented the lessons learned from the relation between networks and consumer electronics, and higher-layer requirements for network infrastructure regarding interoperability and standardization. References [23–26] provide an introduction on the smart grid talking about the problems of the actual grid, the requirements for the future smart grid, the components, benefits, and the entities involved in the development process, framework to classify different types of Energy Management programs and a case against the smart grid. Reference [27] proposed the nanogrid paradigm as an alternative to the Smart Grid top-down approach that adopts the point of view of the grid. Reference [28] illustrated the characteristics of building networks and their relation with the smart grid. Reference [29] presented a set of guiding principles for DR, for each of the domains of utility, customer and energy management.

4.2 Literature on DR Frameworks Directly Applicable to Wholesale Markets

Reference [30] proposed the concept of a Demand Response Exchange in which the demand reduction of interested customers is pooled into the exchange to be bought by various buyers of DR like the retailer,TSO and the DSO. The main idea behind this concept is that the traditional retailer based, TSO based and distributor based DR schemes are only partial solutions to an effective DR program as they focus on benefits of only a particular set of participants. This method aims to optimize the overall benefit DR and moreover reward customers better by allowing them to deal with multiple DR-involved players. The summary of the DR implementations in wholesale markets are presented in [31]. The usefulness of DR in balancing the high penetration of renewables is presented in [32]. In [33], combined day ahead scheduling of energy as well as reserve from both generators and ILs using a probabilistic approach is presented. Authors in reference [34] have devised a complex-bid day-ahead market clearing mechanism in which the DR is scheduled along with the generators energy. Reference [35] proposed a method of congestion management based on the traditionally followed method of congestion management (in decentralized market structures) in which the generators are redispatched after the auction-based dispatch (ignoring network congestion) gives the generation schedules. The generation redispatch may cause additional costs and consequently increase the cost of electricity production. Author's propose the redispatch of demand alongwith generation to handle congestion as an effective solution to mitigate the same in a cost-effective manner.

Reference [36] discussed about the role of a curtailment service provider called Ener NOC bidding DR into the synchronized reserve market in the PJM[1]. The authors in [37] proposed a short-term stochastic SCUC model for scheduling the generating units energy and spinning reserve as well as the reserve provided by the demand response providers (DRPs) simultaneously. Reference [38] presented a non-simplex algorithm to solve an optimization problem in which the objective function is to minimize the production cost of a storage type industry in a scenario in which the electricity price varies over time periods. In reference [39] a short term physically based load model of the demand of a

[1] PJM stands for Pennsylvania New Jersey Maryland

group of air conditioners as a function of outside temperature, air conditioning system parameters and the average thermostat setting. The formulated problem is a linear programming problem, the solution to which provides the optimal schedule of electricity usage given the predetermined electricity price schedule.

References [40], [41] discuss various means for the real-time scheduling of ILs using optimal power flow , dynamic programming, priority-based heuristics inference rules and probabilistic models. Reference [40] presents real-time scheduling of ILs along with generating units by the solution of OPF problem. The conventional OPF is modified to consider a range of ILs at load buses of different power factor and duration of curtailment. Reference [41] propose optimal scheduling of IL portfolios as well as modulating the schedule in the real time owing to influence of load variation and forecasting errors. The customers' requirements are dealt with by fuzzy variables and the optimal or near optimal schedule of ILs are obtained through dynamic programming (DP) search scheme. A priority-based heuristics inference rule (PBHIR) is employed as a regulator in the real-time. The proposed method is called as the model reference adaptive control (MRAC) strategy.

4.3 Literatute on DR Frameworks Directly Applicable to Retail Markets

Most of the papers reported on the DR, points to the retail market implementations. The architechtural needs of the DR esp., for promoting end-consumer participation are addressed in references [42–44].

Scheduling of ILs to achieve a system requirement of hourly curtailments is considered in [86]. This objective alongwith requirements like minimizing payment to the ILs, minimizing frequency of interruptions on ILs form a multi-objective optimization problem which is complex, non-linear and non-continuous problem, solved using binary particle swarm optimization technique. Authors in reference [87], based upon concept of a virtual power plant which comprises of a large number of customers with thermostatically controlled appliances propose an algorithm to determine the optimal control schedules of these appliances or the optimal load reduction bids that an aggregator should present in the electricity market. Reference [47] presented the vision of a home electrical system which consists of solar PV modules, stand-alone energy storage system, electric vehicle, small wind turbine, a bio-fuel burning micro- turbine, set of controllable (TV, lights, etc.) and uncontrollable appliances (HVAC systems). Assumptions involving a day-ahead forecast of the local energy production from wind and solar systems based on weather forecasts, a dynamic pricing signal over the next day and availability of a module to forecast the non-controllable appliances consumption based on the home grid history data are made. The objective is to minimize the cost of energy used over a time horizon. In reference [48] an approach to solve the energy cost minimization problem to find out the optimal hourly consumption level of a number of customers served by a single utility is proposed. In other approaches, the optimization problem is solved centrally. Accordingly, the consumption levels of customers are centrally controlled, that requires high level interactions between the utility company and the customers. Here, the authors propose an Energy Consumption Scheduler (ECS), which can be coupled to the smart meter at each customer premise to solve the optimization problem locally and autonomously using a game theoretic approach.

Authors in [49] proposed an optimization model to adjust the hourly load level of a given consumer in response to hourly electrical prices. The objective is to maximize the utility of the consumer subject to minimum daily energy consumption levels, limits on hourly load levels and ramping limits on such load levels. In [50] a DR model based on Time-of-Use (ToU) and Emergency Demand Response (EDR) methods, using single and multi-period load models is proposed. The paper shows that the demand and load shape can be changed by the SO by implementing such programs. The effect of TOU prices, incentive rates and demand elasticities on shaping the load curve are analyzed. An optimal load scheduling strategy for participating industrial loads is formulated in [51], in a scenario in which DR is implemented through Real-Time Pricing scheme. The objective function is minimization of energy cost (linear function of consumption over periods) with a set of linear constraints such as, amount of electrical energy required to reach production target, bounds on energy consumption in an hour. Rigorous mathematical analysis is done to obtain relationship between cost savings and the parameters like plant installed consumption capacity, plant storage capacity, and terms which describe the structure of the RTP tariff. The approach can be challenged on the grounds of the assumptions made such as that of adequate spare consumption capacity. Authors in [52] proposed a method of scheduling power usage by various appliances in a home in a real-time pricing scenario in which real-time price signals are sent from the utility to the customer end. The proposed method considers scheduling of both-only schedulable as well as both schedulable and real-time appliances over the 24 hours horizon and can be implemented on home area networks (HAN) that may support other applications such as sharing internet access, etc. with a common control channel.

The idea of integrating DR as a source of primary frequency control alongside that from dedicated generators is proposed in [53]. Reference [54] provided an approach called Event-Driven (Emergency) Demand Response in which given a particular type of credible system contingency, particular load reductions are activated, with an objective to restore the operating reserves to a desired level. Authors in [55] proposed the modeling and simulation of commercial building loads especially the HVAC loads in order to estimate the control margin achieved for a certain load control event. The architecture consists of a central controller who calculates the optimal control actions according to the constraints introduced by the retailers/distribution company and the control strategies programmed in the local controllers at the EMS of each consumer building. Reference [56] focused on the business case for a consumer portal demand response functions. Consumer portal acts as a gateway between the utility and the consumer and facilitates many energy service functions including flexible pricing, remote equipment diagnostics, interfacing with building EMS (BEMS), remote energy use monitoring and control,etc. In reference [57], aspects of grid modernization that affect consumer-end activities are addressed, which can be categorized into information and infrastructure, instrument and technology and intelligence and automation. In [58], a framework for jointly controlling regulation from the supply and demand response as a resource for regulation in the real-time imbalance scenario is proposed. The approach is based on a multi-rate model predicting control to capture the varying complementary dynamics of the two resources. In [59] development and validation of physically based load model of electric heating loads is done. A duty cycle model of the demand for evaluating impact of load control strategies is presented in [60]. Reference [61] talks about a discrete time continuous state stochastic model of the temperature of a house having AC/heater regulated by a thermostat. Reference [62] presents a probabilistic model for aggregate storage heating loads taking into account the uncertainty in the external temperature forecasts and the cycling effects of the reduced rate period determined by the utility. In [63] different aggregation techniques for estimation of probability distribution function of thermal and electrical variables of aggregated thermostatically controlled loads whose behaviour is modeled as stochastic differential equation system based on perturbed physical models. A state-queuing model of thermostatically controlled loads is presented to simulate the response of aggregated loads to the electricity prices in [64]

Authors in [65] presented experience on a pilot DR project implemented at Malvik Everk, a DSO in Central-Norway involving 40 household customers. The project involved smart metering, real-time pricing with a time-of-day network tariff and a token provided to the customers indicating peak hours. Specific concepts for DSM involving end-consumers are discussed in references [66–78]. A distribution network perspective of reducing losses by using DR for load shifting is presented in reference [80].

5 Mathematical Perspective of Demand Response

5.1 Optimization

Demand Response is Essentially a Mechanism of Optimal Operation at the Consumer End as Far as the Amount of Energy Consumption is Considered. The Optimal consumption of energy should ensure adequate responsiveness from the demand side, at the same time fulfilling the energy requirements of the consumer. This optimization of energy usage may be motivated by some kind of incentive offered to the consumer from the utility side. For example, a time-varying price signal can be a strong motivation for the consumer to shift loads to the periods of lesser prices or "off-peak" prices. But usually, loads may have a typical time of operation in the entire day. Say for example, a consumer may like the washing machine to be operated in between morning 08:00 till 10:00 hours. In that case, the problem becomes that of a constrained optimization, in which one may have to put limits regarding the time blocks of operation of devices/applications.

The other mathematical perspective of DR is mathematical modeling of various loads, storage equipments, consumer owned generators, etc. Depending on the type of models considered, the optimization problem take the shape of continuous variable optimization, discrete variable optimization, integer variable optimization, binary integer variable optimization and so on. Further, depending on the type of the optimization problem, various optimization techniques have been explored which solve the problem. The literature review section covers a few good literatures [47-56] concerning various DR models. These models cover a wide range of optimization problems of linear and non-linear nature. Various load models have also been presented alongwith associated operational constraints.

In the next subsection, we are going to have a flavor of one such optimization problem in which the energy consumption of a typical residential consumer is optimized. The driving force for such energy scheduling program is a time-varying pricing scheme to which the consumer is subjected.

5.2 Typical DR Framework

The framework under which the DR simulation has been done is depicted in Figure 7. A residential environment is considered for the demonstration purposes. The same can be extended to the Industrial, commercial and other categories of the customers. A single LSE is considered serving a number customers, of which the majority of the customers as in practical cases fall in the residential category. The figure shows the interactions of LSE with a single residential customer. The LSE is considered to be having an internal module which predicts on day d-1 the wholesale market prices of the next day (day d) for all the 24 hours. Also, a forecast of the demand levels for all the hours of day 'd' is also made using the previous consumption data available in the consumption database. Data regarding LSE's own generation if any and other bilateral long term contracts alongwith these short term information is used in an internal optimization program which calculates the 24 hours price data for day 'd', that is to be levied on the customers. The customers are equipped with smart meters which have communication modules in them to receive price signals from the LSE and send overall hourly consumption data to the consumption database. Some of the appliances at the customer ends are assumed to be smart and schedulable appliances, which have an energy controller module fitted onto them. The 24 hours price information is sent to the end use customers by suitable communication means to the smart meters of all the customers on day 'd-1'. The Home Energy Scheduler (HES) available at a residential customer premise uses this information alongwith the input information of the consumption requirements of the

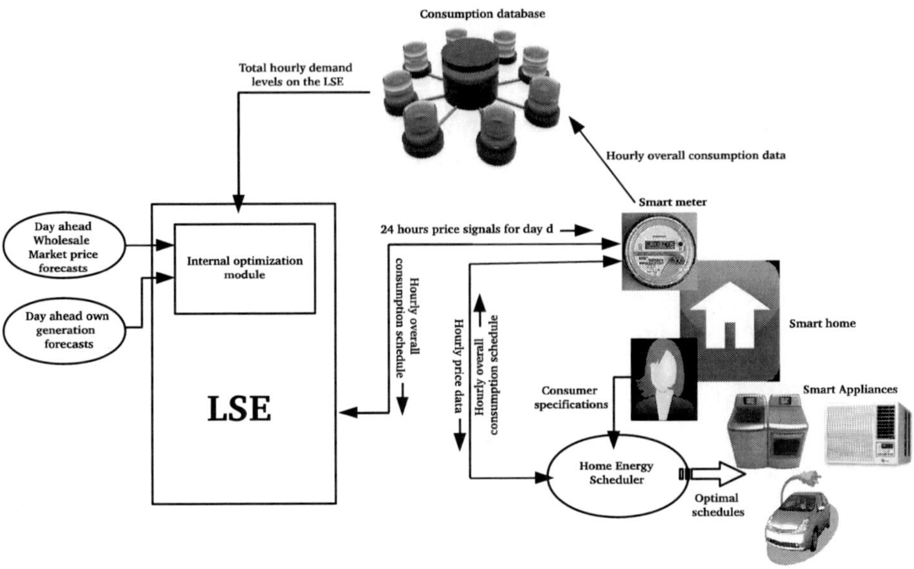

Figure 7. DR framework (considering a residential setup).

concerned customer in order to determine the optimal consumption schedule for day ‘d’ by running a local optimization problem.

A particular residential customer can be assumed to be having a mixture of shedulable (shiftable) and unschedulable appliances. So, as far as consumption scheduling is concerned, the group of schedulable appliances only are of importance. These schedulable appliances are characterized by an operational time band, with a lower limit α_a and an upper limit β_a for each appliance ‘a’ ε A, the set of all appliances. This means the consumption of a particular schedulable appliance can only be scheduled within the hours α_a and β_a, out of which it is in “OFF” state. Further, each appliance is characterized with a predetermined average daily consumption level, which can be determined by the energy controller mounted on the particular appliance. Each appliance a ε A has a maximum power demand level ρ_a^{max} and a minimum standby power level ρ_a^{min} (which in some case may also be equal to zero). This report deals with proposing optimization models for such residential customers, simulating them with a hypothetical residential data and discussing the results obtained.

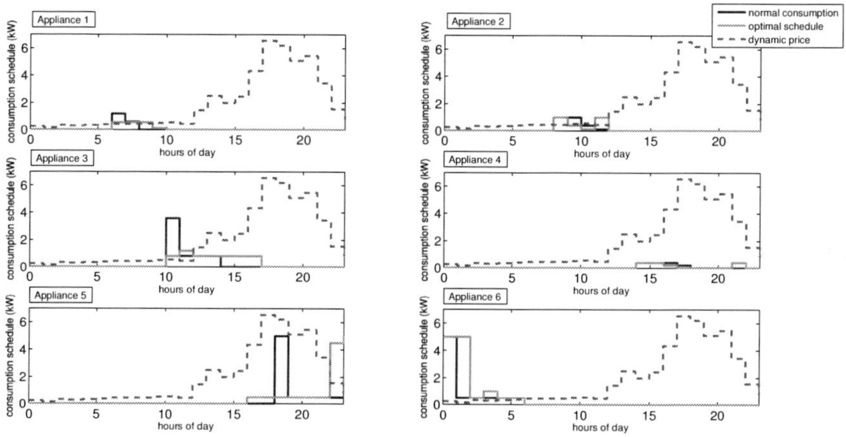

Figure 8. Optimal schedule and normal consumption for appliances 1-6 with model 1.

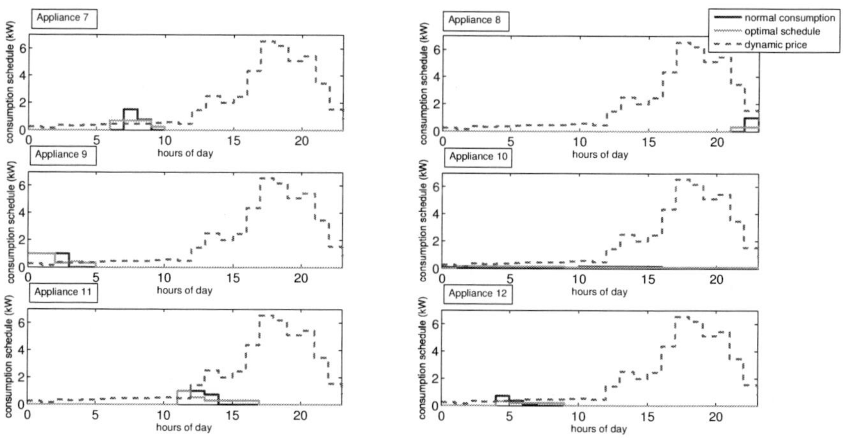

Figure 9. Optimal schedule and normal consumption for appliances 7-12 with model 1.

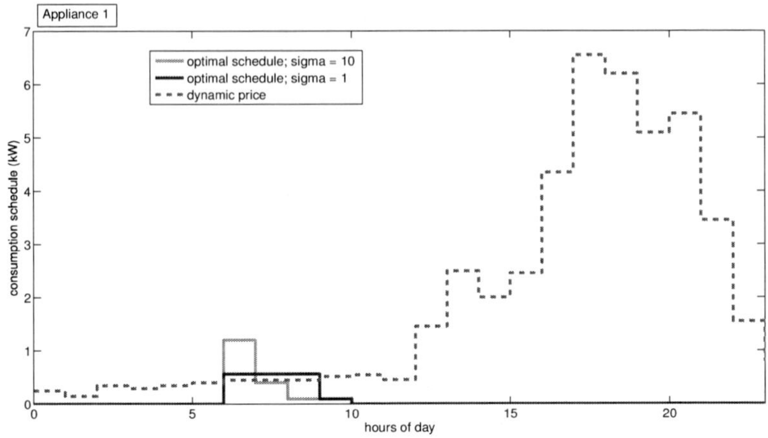

Figure 10. Optimal schedule of appliance 1 with model 2 and waiting factor (σ) = 1 and 10.

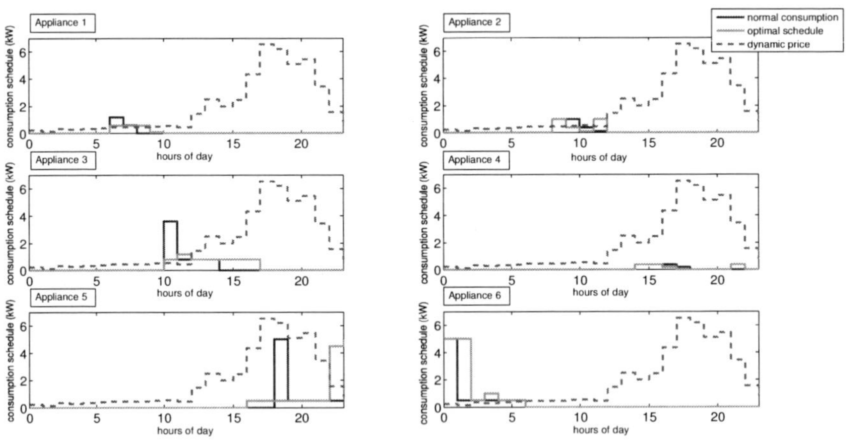

Figure 11. Optimal schedule and normal consumption for appliances 1-6 with model 2 and waiting factors (σ's) = 1.

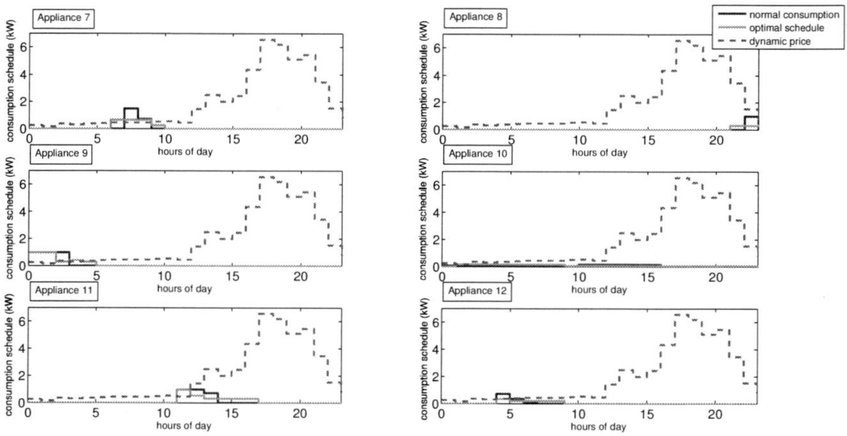

Figure 12. Optimal schedule and normal consumption for appliances 7-12 with model 2 and waiting factors (σ's) = 1.

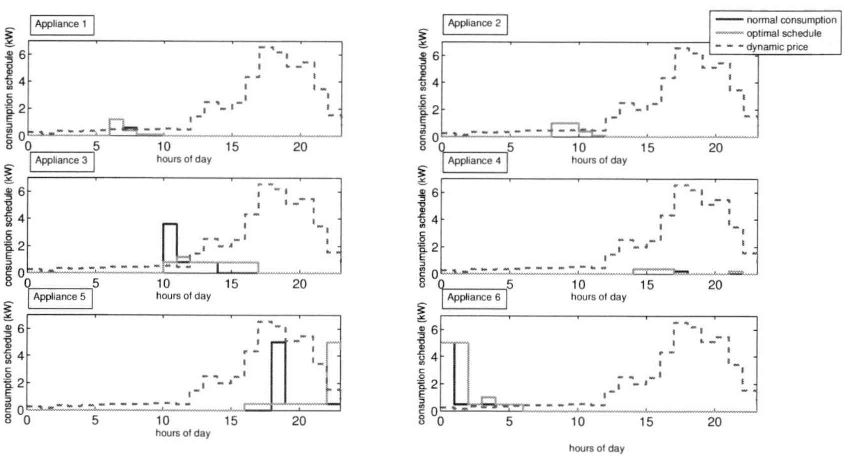

Figure 13. Optimal schedule and normal consumption for appliances 1-6 with model 2.

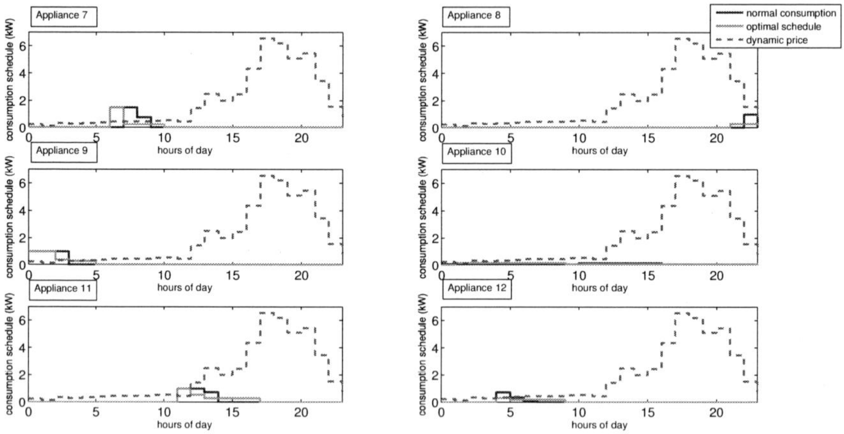

Figure 14. Optimal schedule and normal consumption for appliances 7-12 with model 2.

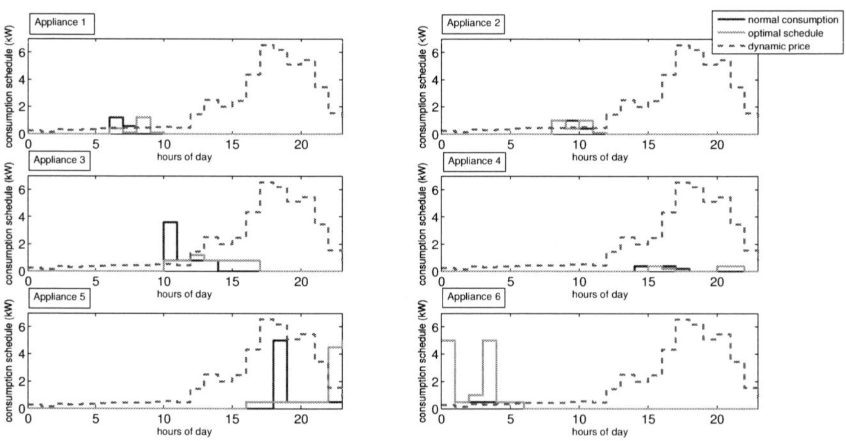

Figure 15. Optimal schedule and normal consumption for appliances 1-6 with model 3.

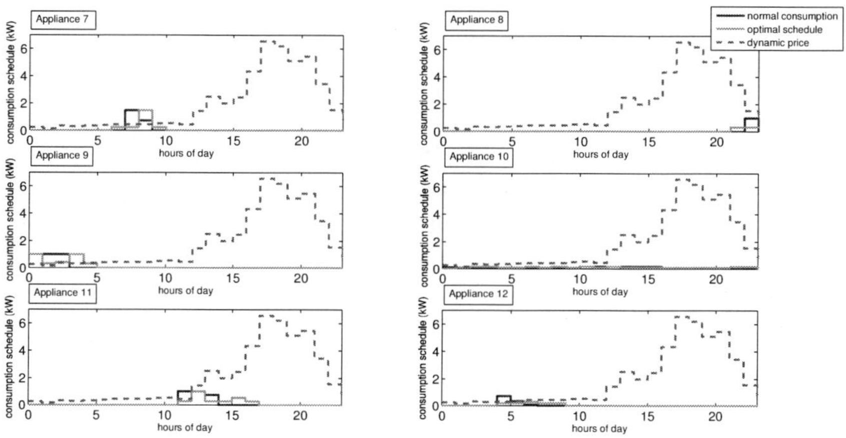

Figure 16. Optimal schedule and normal consumption for appliances 7-12 with model 3.

6 DR Model Formulations

6.1 Model 1

The objective of this model is to find an optimal schedule of the consumption of each shiftable appliance so that the customer's payment to the LSE is minimized, with the given set of 24 hours price signal. Therefore the objective function of the optimization problem can be framed as shown in (1). The constraints of the optimization problem are framed in (2), (3) and (4)

$$Min \sum_{t=1}^{T} \lambda_t * \rho_t \tag{1}$$

where,

$$\rho_t = \sum_{a \varepsilon A} \rho_{at} \tag{2}$$

s.t.

$$\sum_{t=1}^{T} \rho_{at} = E_a \quad \forall \ a \ \varepsilon \ A \tag{3}$$

$$\rho_{at} = 0 \quad for\, t \;\notin\; [\alpha_a, \beta_a] \quad \forall\; a\, \varepsilon\, A \tag{4}$$

$$\rho_a^{min} \leq \rho_{at} \leq \rho_a^{max} \quad \forall\; a\, \varepsilon\, A \tag{5}$$

Constraint (2) relates the total consumption in any time block ‘t’ to the sum of consumption of each and every appliance. Constraint (3) relates the total consumption of an appliance over the entire scheduling horizon to the daily average consumption level which is predetermined. Constraint (4) emphasizes that an appliance consumes no power out of its operational time band. Constraint (5) puts the upper and lower bounds on the consumption levels as depicted by the maximum power levels and minimum or standby levels for each and every appliance.

6.2 Model 2

As argued in model 1, it is quite reasonable to assume that residential customers would like to minimize their bills by scheduling consumption in an optimal way. At the same time depending on the appliance and also the daily timetable of the users they may care about their comfort and getting the work done as soon as possible. But this may be a totally conflicting objective with regard to the former objective of minimizing bills. Therefore, there has to be a compromise between these two objectives to arrive at a feasible optimal schedule. So, this particular model deals in formulating a mathematical model for the comfort of the customer which is actually framed as a cost of waiting. This model is then incorporated alongwith the payment model which becomes the total objective function to be minimized as shown in (6)

$$Min(\sum_{t=1}^{T} \lambda_t * \rho_t + cost \;\; of \;\; waiting) \tag{6}$$

The cost of waiting can be modeled as:

$$cost\ of\ waiting = \sum_{a \varepsilon A} \sum_{t=1}^{T} \theta_{at} * \rho_{at} \tag{7}$$

where,

$$\theta_{at} = \sigma_a^{t-\beta_a} / E_a \tag{8}$$

$\sigma_a \geq 1$ is an adjustable parameter associated with each appliance a ε A. The greater the value of σ_a the greater is the cost of waiting for that particular appliance 'a'. If for any appliance $\sigma_a = 1$, then $\theta_{at} = 1/E_a$ for all the time blocks which signifies that the cost of waiting for the appliance is the same for each and every time block or the user does not lay any importance on the comfort level as far as the particular appliance is concerned. From (8) it is also clear that the more t is closer to α_a the less is the cost of waiting and as $t \rightarrow \beta_a$ the cost of waiting increases for any appliance 'a' if σ_a>1. (7) depicts the overall cost of waiting taken over all the appliances. This model demands an additional input σ_a for each a ε A to the scheduling module of the residential meter.

The set of constraints (2)-(5) apply to this model also.

6.3 Model 3

This model intends to incorporate the aspect of benefit acquired by a residential customer by the usage of electrical energy from the LSE. The objective of this model is maximization of the net benefit the customer obtains which can be mathematically interpreted as:

$$Max(\sum_{t=1}^{T} B(\rho_t) - \sum_{t=1}^{T} \lambda_t * \rho_t - \sum_{t=1}^{T} \sum_{a \varepsilon A} \theta_{at} * \rho_{at}) \tag{9}$$

where,

$$B(\rho_t) = B_{ot} + \lambda_{ot} * (\rho_t - \rho_{ot})(1 + (\rho_t - \rho_{ot})/2E_t\rho_t) \tag{10}$$

The first term in (9) is the benefit function of an individual customer as a function of overall energy consumption in each time block. The mathematical model [84] is shown in (10). The second and third terms are the customer payment to the LSE and the cost of waiting terms respectively. So, the complete expression provides a monetary measure of the net customer benefit. Evidently, this model needs several data pertaining the customer behaviour in order to determine the customer benefit as a function of overall consumption. Nominal price, nominal consumption and the average price elasticity of consumption at each and every time block are such data.

7 Simulation Results

7.1 Data and Solvers

All the models are implemented with a sample empirical hourly day-ahead price data. This dynamic price data for day d is assumed to be provided by the LSE on day d-1 for the optimal scheduling of the residential consumption for the day d. With this data, and the consumption requirement data provided by the customer, the energy scheduler finds the optimal consumption schedule for the day d. Tables 4 and 5 in appendix show typical residential appliance and consumption requirement data pertaining to model 1 and model 2 respectively. Models 1 and 2 are linear programming problems while model 3 has a non-linear component in its objective function in the form of the benefit function. All these models are implemented by using MATLAB optimization toolbox.

7.2 Steps Involved and Consolidated Results

A typical consumption data of each appliance is assumed as the first step. Then, these are the general steps that follow in solving all the three models.

- Processing the raw appliance data to separate them into schedulable and unschedulable appliances.
- Using this processed data, the formulated problem is conditioned as per the required input variables for the particular solver or optimization module.
- The solver is running, which provides the optimal consumption schedule of each appliance over the entire time horizon of 24 hours.
- Cost of unschedulable i.e. the cost of non-shiftable consumption for the hourly price data is calculated.
- Cost of schedulable but unscheduled consumption for the hourly price data as per the typical consumption data is calculated.
- Cost of typical consumption for the hourly price data is then the sum of results in steps 4 and 5.

- Cost of scheduled consumption for the hourly price data is calculated using the optimal schedule obtained in step 3.

- Total optimal consumption cost is calculated as the sum of results in steps 4 and 8.

As seen earlier in section 6 implementation of model 3 requires additional data like nominal price λ_o, nominal consumption ρ_o and average price elasticity E. These data were calculated by generating a set of 7 dynamic (hourly) price data and applying them individually to model 2. The solution of model 2 with these price provided 7 sets of optimal consumption data. The mean of the 7 sets of price data was considered the nominal price data and the mean of the 7 sets of optimal consumption data that so resulted was considered the nominal consumption data. The price elasticities of consumption over 24 hours were calculated using price and consumption data of two consecutive days using the formula:

$$E_t^i = \lambda_t^i/\rho_t^i * (\rho_{t+1}^i - \rho_t^i)/(\lambda_{t+1}^i - \lambda t^i) \quad (11)$$

where, i = 1,2....6 and t = 1,2......24. This results in a set of 6 price elasticity data. The average price elasticity was calculated as the mean of these 6 price elasticity data.

The results thus obtained, when these steps were implemented on the three models are consolidated in Table 2.

Table 2. Implementation results

Cost (in Rs.)↓	Model 1	Model 2	Model 3
Objective function value	58.8904	86.6324	56.4775
Cost of unschedulable consumption	33.3100	33.3100	33.3100
Cost of unscheduled consumption	65.3147	65.3147	65.3147
Cost of scheduled consumption	58.8904	59.5197	62.9466
Cost of typical consumption	98.6247	98.6247	98.6247
Optimal consumption cost	92.2004	92.8279	96.2566
Saving in payment	6.4243	5.7968	2.3681

7.3 Brief Discussion of Results

Figures 8 and 9 show[2] the optimal consumption schedule as obtained by implementation of model 1 using the data in Table 4 in appendix and a typical consumption over 24 hours of day d. It can be clearly seen that all the appliances[3] have been scheduled within their operational time bands in a way that, consumption by a particular appliance is shifted to lower price time blocks within the respective band. This result is intuitive as the optimization objective itself is minimization of the payment made by the residential customer to the utility. As can be observed in Table 2, the optimal consumption cost is Rs. 92.2004 while the typical consumption cost is Rs. 98.6247, resulting in a saving in bill of amount of Rs. 6.4243. Figure 10 shows the effect of varying σ on a particular appliance (here it is appliance 1 i.e. dishwasher). It is observed that when $\sigma =$ 1 for dishwasher the optimal consumption schedule as obtained by model 2 is the same as that obtained by model 1 as shown in Figure 8. When $\sigma = 10$, the optimal consumption schedule is a bit shifted towards the initial hours of the operational time band (which is 7 hours - 10 hours).

On the same lines when model 2 is implemented with $\sigma = 1$ for all the schedulable appliances, the optimal consumption schedule shown in Figures 11 and 12 remains the same as that of model 1. So, we can say, that with $\sigma = 1$ for all the appliances model 2 is a strict cost reduction model.

Now, model 2 is implemented with the appliance data as shown in Table 5 in appendix. Figures 13 and 14 show the results which can be observed to be having slight differences in the optimal consumption schedules of appliances as compared to that of the results in model 1. The appliances are now scheduled with a compromise between minimization of consumption cost and maximization of comfort. The results are listed in Table 2 under the column model 2.

As can be seen, there is an increment in the cost of scheduled consumption as compared to that of model 1 which is due to the shifting of schedules to higher price time blocks in order to maintain comfort levels for some appliances. This results in an optimal consumption cost of Rs. 92.8279 which is more than that of model 1 but still less than the typical consumption cost by Rs. 5.7968 which can be considered as a saving in bills. So, it can be inferred that σ's for some

[2]y-axis is common for both consumption and price

[3]Note: The schedules of only the schedulable appliances have been shown as they are of importance in context to the scheduling algorithm

appliances taking value greater than 1, model 2 is a compromise between cost reduction and comfort maximization.

Figures 15 and 16 show the resulting optimal schedule of appliances when model 3 is implemented with the same set of 24 hours price data.

These plots differ appreciably in terms of optimal schedules of appliances as compared to the previous model results. This is because, in the 3rd model the objective function is maximization of the customer net benefit, instead of only a simple bill minimization or a trade-off between bill minimization and comfort maximization. In comparison to previous models, the cost of scheduled consumption as shown in Table 2 is higher (Rs. 62.9466) and thus the overall optimal consumption cost is also higher as cost of unschedulable consumption is the same as in the previous two models. Nevertheless, the optimal consumption cost of Rs. 96.2566 or the final payment to be made to the LSE is less than the payment of Rs. 98.6247 in case of typical consumption. So, there is a saving of Rs. 2.3681 from the customer's viewpoint.

As observed from the previous discussions, model 1 appears to be the most acceptable one from a residential customer viewpoint when there is no concern regarding the comfort of the customer. Model 2 incorporates the aspect of customer comfort and thus adds a trade-off between the two conflicting objectives of comfort maximization and payment/bill minimization, as a result of which the customer payment to the LSE increased as compared to that in model 1. This increment in bill can be considered as the cost of the desired comfort level of the customer. So, this model seems to be a very comprehensive one from the viewpoint of a customer, as by changing the waiting parameters (σ's) the desired comfort levels can be changed.

In contrast, model 3 appears to be a more comprehensive model when looked from the viewpoint of the LSE as the results show some changes in consumption patterns as compared to a typical consumption pattern. Obviously, some demand levels have shifted to the lower price blocks or otherwise the payment of the customer would not have been less than that when the consumption pattern is a typical one. So, the first objective of shifting consumption levels to low price blocks is achieved. Secondly, the loss of revenue for the LSE is reduced in model 3 as the payment from optimal consumption is greater than that of the other two models. And, thirdly, the LSE is able to achieve customer satisfaction by allowing them to choose their desired comfort levels and helping them

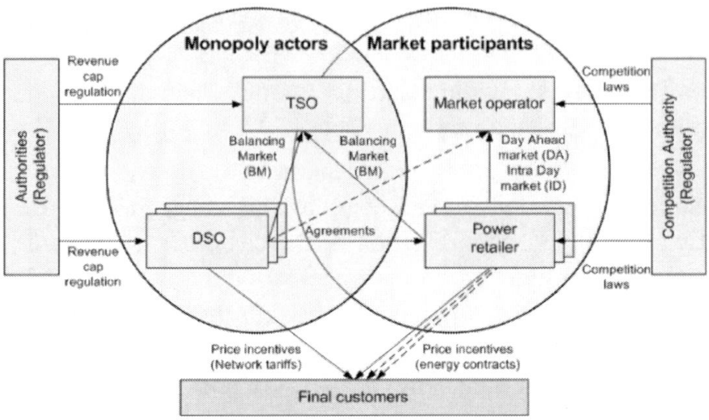

Figure 17. Norwegian market structure.

achieving savings in electricity bills.

8 DR Pilots and Challenges in Practical Implementation

8.1 Pilots and Implementations

DR pilots and implementations are being carried out in many parts of the world. References [65] provide some ongoing and futuristic DR implementations and pilots across various regions. To make the concept of DR pilots more clear, we describe in detail a particular pilot project that was carried out in Norway [65]. This project exemplifies DR pilots very effectively. The pilot study described in [65] was adapted for the residential customers in the Nordic market. Residential customers in Norway (consume around 15000 kWh per customer per year) have the highest electricity consumption in Europe. The Norwegian market structure is depicted in Figure 17. The distribution system operator (DSO) and the power retailer are two separate entities in the Norwegian deregulated power system . So all customers have two separate contracts: one tariff for grid usage and one contract for energy usage. Network tariff design is vested with

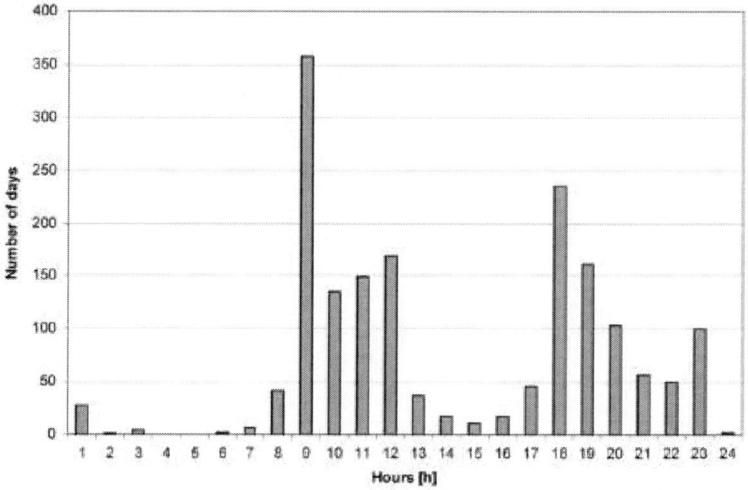

Figure 18. Histogram of the occurrence of maximum spot prices for Trondheim 2005-2009 (data source: NordPool.)

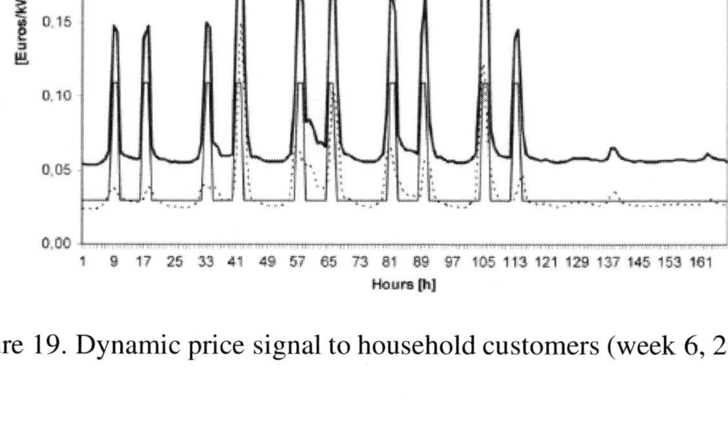

Figure 19. Dynamic price signal to household customers (week 6, 2007.)

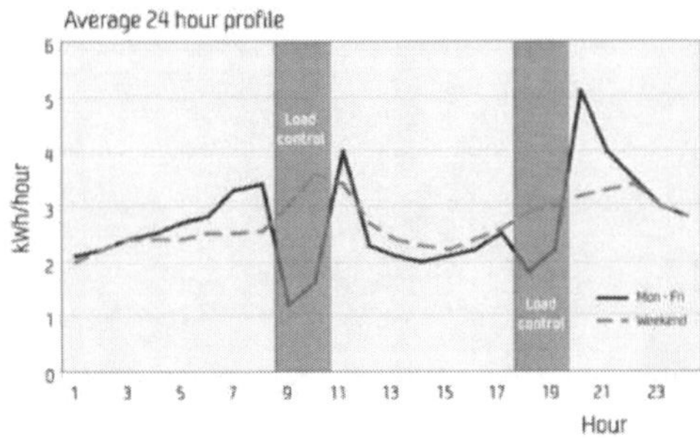

Figure 20. Load profile for a household customer with hot water space heating system and RLC.

the monopoly regulation. And the costs for the DSO can be divided between the two parts. The traditional network tariff to a household customer will therefore be: The energy contract which is between the power retailer and the final customer is amenable to flexibilities at the retailer's discretion. Various options, for example fixed energy price for a defined period of time or a spot related price can be exercised by the retailers. The end customers also receive some price incentives (network tariffs and energy contracts) from the monopoly and market side respectively.

The motivation for the pilot was to test if customers with an energy contract reflecting the spot price on an hourly basis, hourly metering and hourly settlement of their consumption, benefit from reduced consumption in peak hours.

For that an initial survey of the load curves and the spot prices of electricity for Trondheim (Central-Norway) was done. Hourly spot prices for Trondheim (Central-Norway) from 1 January 2005 to 9 October 2009 were used. The total number of days in this period was 1743.

Figure 18 shows a histogram showing the hours when the maximum spot prices occur. The repeating trend of peak load periods with high spot prices during the day can be seen. Also it is observed that the maximum price occurred in hour 9 for 358 days (20.5%) and in hour 18 for 235 days (13.5%). Further it is observed that the peak load periods and high spot prices occur on an irregular basis, but mainly in the same hours of the day (morning and afternoon).

Based on these findings the Time of Day network tariff devised had the following structure:

In what follows are excerpts from the reference [65] itself about the pilot.

"The pilot study was implemented at Malvik Everk, a DSO in Central-Norway. The invitation to customers was presented in an article in the local newspaper and via letters sent by mail. Prior to the pilot study all the DSO's household customers had installed Automated Meter Reading (AMR) for weekly metering of their electricity con- sumption. In relation to this pilot study 40 household customers were given hourly metering of their electricity consumption with use of existing AMR technology. Ten percent of the customers in the pilot study had a hot water space heating system with electrical boiler with installed ca- pacity of 12-15 kW. The rest of the customers had standard electrical water heaters with installed capacities of 2-3 kW. Extra installation costs related to the Remote Load Control (RLC) ranged from 80-375 Euros per customer. The installation costs consisted of labor and material costs and were covered by the DSO. Two information meetings were performed during the pilot study. At these meetings the customers were informed about the different aspects of the pilot study (price signals, peak hours, load control, etc.), possibilities for changing demand and the benefits of DR. The customers also got a web site with information about their network costs with the Time-of-Day network tariff, compared with the traditional network tariff. The DSO had to get an exemption from today's regulations to be able to offer the ToD network tariff used in the pilot study. Therefore the ToD network tariff is not a tariff that can be offered to customers at a large scale today. The duration of this pilot study was one year.

Each household in the pilot study was equipped with three small tokens, the "El-buttons," to be placed on dishwashers, washing machines, etc., to remind the households to avoid usage of these energy consuming appliances in the predefined peak load periods. The token represents a watch with 12/24 hours, where the leftmost sector represents the morning peak hours (08:00–10:00) and the lower sector represents the afternoon peak hours.

The household customers in the pilot study got a ToD network tariff which stimulated load shifting. RLC via the AMR system was offered as an aid to reduce load and costs in the predefined peak load periods. Additionally, they were advised to have an energy contract with the spot price on an hourly basis. Figure 19 shows an example dynamic price signal offered to the customers."

Figure 20 shows the load shifting for a household customer with hot water space heating system and RLC of the electrical boiler. The customer benefited by shifting consumption from high price to the lower price hours.

By aggregating the response over 50% of Norwegian households an estimate of DR potential was made to be at 1000 MWh/h that was around 4.2% of peak load demand in Norway. Another inference drawn by the authors from this pilot is that load reduction of this size can be utilized as a product in the day ahead market. This can help in reducing actual price of electricity as a proactive measure instead of reacting to the spot prices. The latter may also cause imbalance issues necessitating activation of balancing resources, creating additional costs.

Aggregated residential DR, industrial and commercial DR are in principle eligible as wholesale market products. [83] In the context of energy and ancillary service markets facilitated by the ISOs/RTOs, depending on the ISO/RTO market design and operational standards, the market products DR can offer may include some or all of the following:

- capacity (installed capacity or unit commitment avail ability in some markets);
- energy (day-ahead, real-time balancing);
- reserve A/S, which may include some or all of the following:
  - regulating reserve;
  - spinning reserve/responsive reserve;
  - supplemental reserve/nonspinning reserve.
- Out-of-market (or out of merit order) deployment of DR may occur under emergency conditions.

Depending on the specific ISO/RTO design, the DR products may be offered into the markets by market participants like LSEs, CSPs, aggregators, etc.

In what follows is the description given in [83] about the different DR products and rules in Pennsylvania-Jersey-Maryland (PJM), one of the several US ISOs.

- Economic load response, which may be provided by agent PJM members or CSPs. It is an economic program that uses the LMP as a trigger. Consumption under this program is curtailed when $LMP \geq \$75$

- Energy dispatched out of DR sold as capacity or A/S. The energy dispatched out of capacity or A/S DR is paid the real-time LMP. The compensation is split between the CSP and the LSE as follows:

  - compensation to CSP: zonal LMP minus retail rate (generation and transmission portion)
  - compensation to LSE: retail rate DR participation in PJM's capacity market enables LSEs and CSPs to offer DR as a capacity product and receive capacity credit for the MW quantity of DR that clears the capacity market.

For DR to participate in PJM's A/S markets, the following rules apply.

- For DR participation in PJM's synchronized reserve (spin- ning reserve) market:

  - DR must be able to provide metering data at no less than a 1-min scan rate.
  - DR participation in synchronized reserve market is lim- ited to 25% of the synchronized reserve requirement in each zone.
  - Mandatory training requirements exist for CSPs bidding

- DR in the synchronized reserve market. For DR participation in PJM's regulation market:

  - DR bidding regulation must meet all real-time telemetry requirements like a generator.
  - DR participation in the regulation market is limited to 25% of the regulation requirement.
  - There are mandatory training requirements for CSPs bidding DR in the regulation market.

Such DR market products are offered by other US ISOs like NYISO, CAISO, ISO-NE and so on.

8.2 Challenges

There are several challenges in implementing DR in a smart grid paradigm. One of the most important challenges in DR implementation is making the end consumer aware of the benefits and implementation issues. Broad market analysis coupled with studies on product design can ensure fair acceptance of the technology by the consumer side to a great extent. Regulators, policy-makers and financial institutions should also participate in DR educational programs. There is always a question of uncertainty over the real potential of new technologies and tariff-designs until and unless robust empirical evidence of the their performance and economics are available over time on a system-wide basis. Moreover, the utility industry may not be able to support a financial roll out to fund new technologies without the help of government incentive programs. The regulators will always warrant extensive proof before authorizing investments [81,82]. Further replacement of older infrastructure for example replacement of electromechanical electricity meters by new smart meters and other smart technologies may pose a problem for the utilities. Such early replacement of equipments may pose financial burdens to the end consumers. Many countries still lack policies, standards and required Information and Communication Technology (ICT) infrastructure for DR implementation. Limited experience of the distribution utilities with ICT technologies pose further difficulties in DR implementation. To alleviate this problem, distribution companies must coordinate with leading ICT companies in acquiring relevant knowledge and technology know how. Development of standards and protocols regarding the design and operation of AMI are still in the process of maturity and thus investment made in such a situation is always subject to the risk of obsolescence. A cost benefit analysis considering these issues with an estimate of the cost of the equipments the customer will deploy to automate response has to be made to reduce the investment risk. A proper rate should be designed not only to ensure that the investment made in the new technology is recovered in fair amount of time, but also to be acceptable to every customer class.

9 Future of DR Programs and Research Opportunities

Some areas for research in DR are been identified to promote them as Smart DR.

- Study the operation of DR programs with relevant standards to increase penetration of such programs to the distribution levels
- Standardization of technical and commercial protocols and hardware to allow easy installation of enabling devices with plug-and-play capabilities
- Institutional and business models of DR for empowerment of end-consumer
- Technology-society interface and policy related studies
- Study the operation of DR in parallel with the grid and islanded under various faults (With and without communication infrastructure)
- Define and develop control strategies under interoperability standards environment to ensure efficient, reliable, and economic operation and management of DR
- Define appropriate protection and grounding policies to assure safety, fault detection, separation, and islanded operation under advanced communication infrastructure
- Exploring the impact on the development of electricity network infrastructures, including quantification of the benefits of DR, to the overall network, and to the reinforcement and replacement strategy of the aging electricity infrastructure
- Identify and develop the required telecommunication infrastructures and protocols to enable faster information exchanges
- Determine the economic benefits of DR operation and propose systematic methods to quantify them under interoperability standards environment (for coordination with other DR)
- Simulate and demonstrate standards based DR operation on laboratory scales.

- Developing DER models plus steady-state and dynamic analysis tools for enabling simulation of LV asymmetrical, inverter dominated DR performance
- Islanded and interconnected operating philosophies under interoperability standards environment
- Developing control algorithms, both hierarchical and distributed (agent based)- for DR operations
- Executing extensive field trials of alternative control strategies in actual installations, with experimental validation of various DR architectures in interconnected and islanded modes, and during transition testing of power electronics components and interfaces and of alternative control strategies, communication protocols, etc.
- Local blackstart strategies using communication infrastructure
- Definitions of DER interface response and intelligence requirements
- Grounding and protection schemes for real-time operations
- Advanced methods for quantification of reliability benefits
- Laboratory DR of various complexities and functionalities- like demonstrating coordination between DR Applications
- Investigation of new DER controllers to provide effective and efficient operation of DR
- Development of alternative control strategies using next generation information and communications technology
- Creation of alternative network designs, including application of modern protection methods, modern solid-state interfaces, and operation at variable frequencies
- Technical and commercial integration of multiple DR, including interface of several DR with upstream distribution management systems, plus operation of decentralized markets for energy and ancillary services using interoperability standards

- Advanced communication infrastructure and studying the impact on power system operation, including benefits quantification of DR at state, regional, and national levels of reliability improvements, reduction of network losses, environmental benefits, etc.

10 Role of Standards

Open and interoperability standards will definitely enables the inter and intra communications in the DR by promoting the vendor neutrality and mapping with the energy management systems. It also enables the plug-and-play capabilities to a DR. The standards also provide a common portfolio for a variety of generation technologies, and coordination between loads. Absence of the standards slows down the proliferation of various renewable energy technologies integration to the DR schemes. OpenADR, IEC PC118, IEEE P1547, IEEE 929-2000 and IEEE 519 standards are available for DR, grid-connected DG/DER systems [112, 113], which includes:

- DR interface design for utility
- DR communications from utility to the consumers
- Impacts of voltage, frequency, power quality,
- Inclusion of single point of common coupling (PCC) and multiple PCCs,
- Protection schemes and modifications,
- Monitoring, information exchange and control,
- Understanding load requirements of the customer,
- Knowing the characteristics of the DG/DER,
- Identifying steady state and transient conditions,
- Understanding interactions between machines,
- Reserve margins, load shedding, demand response,
- Cold load pickup, additional equipment requirements, and

- Additional functionality associated with inverters.

Similar standards need to be developed for various “domains” in the DR (isolated/grid connected or rural/urban) as well, which will help in daily operation of such systems and incorporates smartness to the system. The purpose of these standards is also to build semantic and syntactic understanding between the various applications under the DR. This advances the application functioning by incorporating smartness (coordination with other applications) in their computations and decision analysis. Many groups like IEEE, IEC, MultiSpeak etc., are working towards developing the open and interoperability standards. It is recommended the support of international standards for large-scale deployment of DR to sustain. This will do inturn definitely benefit both the utility and the consumers.

10.1 OpenADR

Demand Response (DR) programs help utilities maintain grid reliability and enable customers to realize significant value. Unfortunately, existing propriety solutions add unnecessary cost and complexity. The key feature of the demand response under the smart grid paradigm is the availability of the non-proprietary, open standardized DR interface that allows electricity providers to communicate DR signals directly to existing customers using a common language and existing communications such as the Internet. This can be achieved by using the Open Automated Demand Response (OpenADR). OpenADR Alliance is promoting the OpenADR and it was created to standardize, automate and simplify DR to enable utilities to cost-effectively meet growing energy demand, and customers to control their energy future [114].

The history of the OpenADR development is as follows-

$\Downarrow$ 2002 - Research begun at LBNL Demand Response Research Center (DRRC) into automated DR

$\Downarrow$ 2003 – 2006 Numerous pilots and field trials

$\Downarrow$ 2007 - Present – Commercialization and use of OpenADR in PG&E, SCE, and SDG&E DR programs to automate DR programs with primarily Commercial and Industrial customers. Number of DR programs utilizing OpenADR continues to expand.

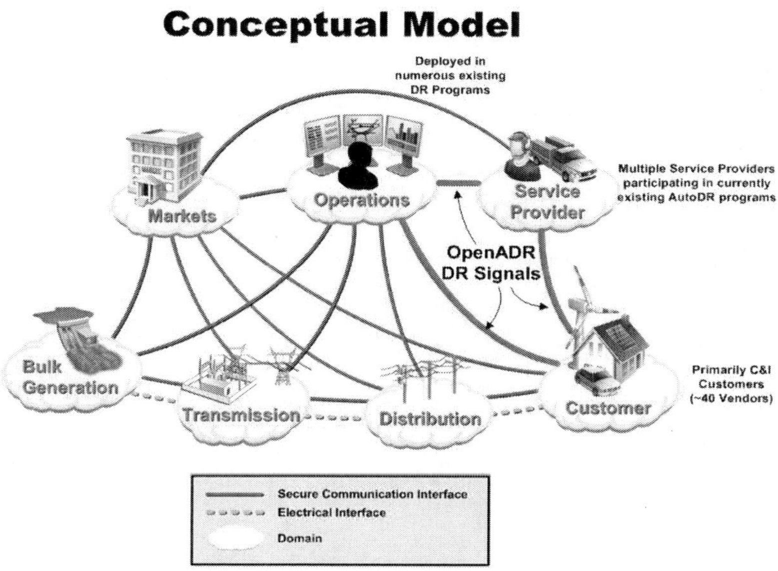

Figure 21. Conceptual model representing the OpenADR signals.

Table 3. Benefits of openADR

Utilities and Energy Service Providers	Commercial, Industrial and Residential Customers	Equipment Manufactures and System Integrators
• Increase grid reliability – more predictable load control • Defer capital investment – an alternative to new generation • Achieve DR goals – comply with state policies & targets	• Mitigate CPP events – minimize dynamic pricing impact • Monetize discretionary loads – utilize available capacity • Maximize utility incentives – subsidize new system investments	• Differentiated products – compete against proprietary systems • Reduce complexity – quicker, easier installation & operation • Enhanced customer service – help customers control costs

⇓ Q1/2007 – Standardization effort for OpenADR begins at DRRC with Industry Consortium drafting specification

⇓ Q1/2009 – First official draft of OpenADR specification is published by California Energy Commission. Work is donated to UCAIug and OASIS

⇓ 2009 – OpenADR taskforce formed within UCAIug

⇓ 2009 – Energy Interoperation Technical Committee formed within OASIS

⇓ 2009 – NIST develops SmartGrid Roadmap in which OpenADR is identified as one of the original 16 low hanging fruit.

⇓ Q3/2009 – NIST forms Priority Action Plan (PAP) 09 for standardizing DR signals. OpenADR taskforce works closely with NAESB to develop requirements for PAP 09

⇓ 2010 – UCAIug focus on OpenADR SRS and SD

⇓ 2012- OpenADR 2.0

10.2 IEC PC118

IEC PC118 is the project committee formed by the International Electrotechnical Commission [116] to standardize the smart grid interfaces with the objectives as

- Standardization in the field of information exchange for demand response and in connecting demand side equipment and/or systems into the smart grid.
- More focused on International perspective on the Smart Grid User Interface (SGUI) to enable standardized DR operations.

Energy service providers and facility service consumers are the key components in any DR program. The conceptual architecture representing the actors and the corresponding information flows is depicted in Figure 22.

As depicted- Loads, storage, generators and meters fall into the facility service consumers category. External service providers like wholesale markets,

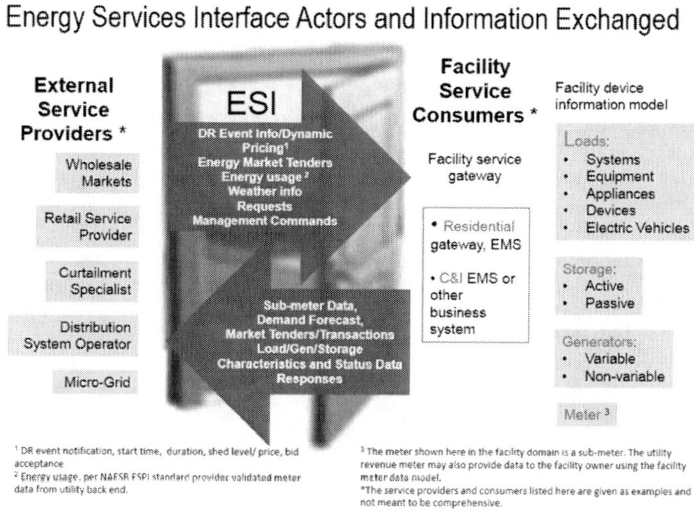

Figure 22. Energy Services Interface Actors and Information Exchanged for DR.

retail service provider, curtailment specialist, distribution system operator, and Micro-grids wants to communicate with the consumers for different DR actions through a common standardize interface. The standardization work is carried out under the guidance of the IEC PC118.

Conclusion

The chapter covered several important aspects of Demand Response (DR). Beginning with a brief background of the topic, a vast literature review is presented for a complete exposure to the reader.

From the review it is clear that DR covers both the wholesale (transmission) and retail (distribution) sectors of the power domain. Further a mathematical perspective is put forward to bring out the underlying technical considerations in practical implementation of DR programs. Three different DR models for a price based DR program were presented. With these sample problems, simulation results showing effective load management and control are discussed.

Practical implementation challenges are brought out with studies of DR pilots and actual implementations. DR standardization is the other important issue that has been covered. The chapter closes with remarks on DR standardization and future scope in this field with research opportunities identification.

Appendix

Table 4. Residential appliance data for model 1

Appliance	type	no	ρ_a^{max}(kW)	ρ_a^{min}(kW)	α	β	E_a(kWh)
Dishwasher	1	1	1.20	0.100	7	10	1.50
Air conditioner	1	1	1.00	0.100	9	12	2.50
Clothdryer	1	1	4.00	0.800	11	17	1.50
Washing machine	1	1	0.40	0.000	15	22	3.50
PHEV	1	1	5.00	0.500	17	24	2.50
PHEV	1	1	5.00	0.500	1	6	2.50
Water heater	1	1	1.50	0.250	7	10	1.50
Room heater	1	1	1.00	0.300	22	24	1.00
Room heater	1	1	1.00	0.300	1	5	3.00
Refrigerator	1	1	0.15	0.075	2	23	15.00
Air conditioner	1	1	1.00	0.300	12	17	2.75
Pump motor	1	1	0.75	0.200	5	9	1.50
Lamps	0	5	0.10	0.000	1	24	25.00
Tube Lights	0	5	0.04	0.000	1	24	25.00
Electric Iron	0	1	1.00	0.000	1	24	1.00
Toaster	0	1	0.75	0.000	1	24	0.75
Television	0	1	0.20	0.000	1	24	7.00
Fans	0	5	0.10	0.000	1	24	10.00

type '1' = schedulable appliance; type'0' = unschedulable appliance

Table 5. Residential appliance data for model 2 and 3

Appliance	type	no	ρ_a^{max}	ρ_a^{min}	α	β	σ	E_a(kWh)
Dishwasher	1	1	1.20	0.100	7	10	10.0	1.50
Air conditioner	1	1	1.00	0.100	9	12	10.0	2.50
Clothdryer	1	1	4.00	0.800	11	17	1.0	1.50
Washing machine	1	1	0.40	0.000	15	22	10.0	3.50
PHEV	1	1	5.00	0.500	17	24	10.0	2.50
PHEV	1	1	5.00	0.500	1	6	1.0	2.50
Water heater	1	1	1.50	0.250	7	10	30.0	1.50
Room heater	1	1	1.00	0.300	22	24	1.0	1.00
Room heater	1	1	1.00	0.300	1	5	10.0	3.00
Refrigerator	1	1	0.15	0.075	2	23	1.5	15.00
Air conditioner	1	1	1.00	0.300	12	17	1.7	2.75
Pump motor	1	1	0.75	0.200	5	9	1.5	1.50
Lamps	0	5	0.10	0.000	1	24	1.0	25.00
Tube Lights	0	5	0.04	0.000	1	24	1.0	25.00
Electric Iron	0	1	1.00	0.000	1	24	1.0	1.00
Toaster	0	1	0.75	0.000	1	24	1.0	0.75
Television	0	1	0.20	0.000	1	24	1.0	7.00
Fans	0	5	0.10	0.000	1	24	1.0	10.00

References

[1] H. Fraser, "The importance of an active demand side in the electricity industry," *The Electricity Journal, Elsevier*, Volume 14, Issue 9, pp. 52-73, November 2001.

[2] G. Strbac, "Demand side management: Benefits and challenges," *Energy Policy, Elsevier*, Volume 36, Issue 12, pp. 4419-4426, December 2008.

[3] S. Braithwait, "Behavior modification," *Power and Energy Magazine, IEEE*, Vol. 8, Issue 3, pp. 36-45, May-June 2010.

[4] A. K. David, Y. Z. Li, "A comparison of system response for different types of real time pricing," in *Proc.of IET International Conference on Advances in Power System Control, Operation and Management*, 1991, vol. 1, pp. 385-390.

[5] D. Yang and Y. Chen , "Demand response and market performance in power economics:," in *Proc. of IEEE Power and Energy Society General Meeting*, 2009, pp. 1-6.

[6] G. Strbac, E. D. Farmer and B. J. Cory, "Framework for the incorporation of demand-side in a competitive electricity market," in *Proc. of IEE Generation, Transmission and Distribution*, vol. 143, no. 3, pp. 232-237, May 1996.

[7] T. T. Nguyen and A. Yousefi, "Multi-objective demand response allocation in restructured energy market," in *Proc. of IEEE Innovative Smart Grid Technologies (ISGT)*, 2011, pp. 1-8.

[8] M. Behrangrad, H. Sugihara and T. Funaki, "Analysing the system effects of optimal demand response utilization for reserve procurement and peak clipping," in *Proc. of IEEE Power and Energy Society General Meeting*, 2010, pp. 1-7.

[9] P. Faria, Z. Vale, J. Soares, J. Ferreira, "Demand response management in power systems using a particle swarm optimization approach," *IEEE Intelligent Systems*, vol. PP, no. 99, pp. 1-9, Apr. 2011.

[10] J. A. Short, D. G. Infield, and L. L. Freris,"Stabilization of grid frequency through dynamic demand control," *IEEE Trans. on Power Syst.*, vol. 22, no. 3, pp. 1284-1292, Aug. 2007.

[11] A. J. Roscoe and G. Cult , "Supporting high penetrations of renewable generation via implementation of real-time electricity pricing and demand response," *IET Renewable Power Generation*, vol. 4, no. 4, pp. 369-382, July 2010.

[12] T. H. Yoo, Hun-Gyu Kwon, H. C. Lee, C. H. Rhee, Y. T. Yoon and J. K. Park , "Development of reliability based demand response program in Korea.," in *Proc. of IEEE Innovative Smart Grid Technologies (ISGT)*, 2011, pp. 1-6.

[13] J. Medina, N. Mueller and I. Roytelman , "Demand response and distribution grid operations: Opportunities and challeneges," *IEEE Trans. on Smart Grid*, vol. 1, no. 2, pp. 193-198, Sept. 2010.

[14] S. Mohagheghi, J. Stoupis, Z. Wang, Z. Li and H. Kazemzadeh , "Demand response architecture- Integration into the distribution management system," in *Proc. of IEEE First International Conference on Smart Grid Communications (SmartGridComm)*, 2010, pp. 501-506.

[15] D. T. Nguyen , "Demand Response for Domestic and Small Business Consumers : A New Challenge," in *Proc. of IEEE Transmission and Distribution Conference and Exposition*, 2010, pp. 1-7.

[16] M. Negnevitsky, T. D. Nguyen and M. de Groot , "Novel business models for demand response exchange," in *Proc. of IEEE Power and Energy Society General Meeting*, 2010, pp. 1-7.

[17] S. Lu, N. Samaan, R. Diao, M. Elizondo, C. Jin, E. Mayhorn, Y. Zhang and H. Kirkham, "Centralized and decentralized control for demand response," in *Proc. of IEEE Innovative Smart Grid Technologies (ISGT)*, 2011, pp. 1-8.

[18] S. A. Pourmousavi and M. H. Nehrir, "Demand response for smart microgrid: Initial results," in *Proc. of IEEE Innovative Smart Grid Technologies (ISGT)*, 2011, pp. 1-6.

[19] N. Mahmoudi-Kohan and M. P. Moghaddam, "Improving WFA K-means technique for demand response programs applications," in *Proc. of IEEE Power and Energy Society General Meeting*, 2009, pp. 1-5.

[20] A. Mutanen, M. Ruska, S. Repo and P. Jarventausta, "Customer classification and load profiling method for distribution systems," *IEEE Trans. on Power Delivery*, vol. 26, no. 3, pp. 1755-1763, June 2011.

[21] S. Valero, M. Ortiz, C. Senabre, C. Alvarez, F. J. G. Franco and A. Gabaldon, "Methods for customer and demand response poilicies selection in new electricity markets," *IET, Generation Transmission and Distribution*, vol. 1, no. 1, pp. 104-110, Jan. 2007.

[22] B. Nordman, "Networks in buildings: which path forward?," *Proceedings of the ACEEE Summer Study on Energy Efficiency in Buildings*, August 2008.

[23] W. Frye, "Smart grid: transforming the electricity system to meet future demand and reduce greenhouse gas emissions," *Cisco Internet Business Solutions Group*, November 2008.

[24] M. A. Piette, S. Kiliccote, and G. Ghatikar, "Linking Continuous Energy Management and Open Automated Demand Response," *Report LBNL-1361E, Proceedings of the Grid Interop Forum, Atlanta*, November 2008

[25] A. Ipakchi, F Albuyeh, "Grid of the Future," *Power and Energy Magazine, IEEE*, Vol. 7, Issue 2, pp. 52-62, April 2009.

[26] B. Nordman, "The Case Against the Smart Grid," *presentation and video available on YouTube*, October 2009.

[27] B. Nordman "Nanogrids: Evolving our electricity systems from the bottom up," *Environmental Energy Technologies Division - Lawrence Berkeley National Laboratory*, May. 2010.

[28] B. Nordman "Beyond the Smart Grid: Building Networks," *Environmental Energy Technologies Division - Lawrence Berkeley National Laboratory*, May. 2010.

[29] K. Wacks, "Open Energy Management Architecture," *iHomes & Buildings, CABA*, pp. 16-19, Spring 2010.

[30] Duy Thanh Nguyen, Michael Negnevitsky and Martin de Groot, "Pool-based demand response exchange– concept and modeling," *IEEE Trans. on Power Syst.*, vol. 26, no. 3, pp. 1677-1685, Aug. 2011.

[31] M.H. Albadi, E.F. El-Saadany, "A summary of demand response in electricity markets," *Electric Power Systems Research, Elsevier*, Volume 78, Issue 11, pp. 1989-1996, November 2008.

[32] I. Stadler, "Power grid balancing of energy systems with high renewable energy penetration by demand response," *Utilities Policy, Elsevier*, Vol. 16(2), pp. 90-98, June 2008.

[33] J. Bai, H. B. Gooi, L. M. Xia, G. Strbac, and B. Venkatesh, "A prob- abilistic reserve market incorporating interruptible load," *IEEE Trans. Power Syst.*, vol. 21, no. 3, pp. 1079-1087, Aug. 2006.

[34] Chua-Liang Su and Daniel Kirschen, "Quantifying the effect of demand response on electricity markets," *IEEE Trans. on Power Syst.*, vol. 24, no. 3, pp. 1199-1207, Aug. 2009.

[35] E. Shayesteh, M. Parsa Moghaddam, S. Taherynejhad and M. K. Sheikh-EL-Eslami "Congestion management using demand response programs in power market," in *Proc. of IEEE Power and Energy Society General Meeting Conversion and Delivery of Electrical Energy in the 21st Century*, 2008, pp. 1-8.

[36] K. Schisler, T. Sick and K. Brief , "Role of demand response in ancillary services markets," in *Proc. of IEEE Transmission and Distribution Conference and Exposition* , 2008, pp. 1-3.

[37] M. Parvania and M. Fotuhi-Firuzabad, "Demand response scheduling by stochastic SCUC," *IEEE Trans. on Smart Grid*, vol. 1, no. 1, pp. 89-98, June 2010.

[38] B. Daryanian, R. E. Bohn and R. D. Tabors, "Optimal demand-side response to electricity spot prices for storage-type customers," *IEEE Trans. on Power Syst.*, vol. 4, no. 3, pp. 897-903, Aug. 1989.

[39] T. M. Calloway and C. W. Brice, "Physically based model of demand with applications to load management assessment and load forecasting," *IEEE Trans. on Power App. Syst.*, vol. PAS 101, no. 12, pp. 4625-4631, Dec. 1982

[40] S. Majumdar, D. Chattopadhyay, and J. Parikh, "Interruptible load management using optimal power flow analysis," *IEEE Trans. Power Syst.*, vol. 11, no. 2, pp. 715-720, May 1996.

[41] K. Y. Huang, H.-C. Chin, and Y.-C. Huang, "A model reference adaptive control strategy for interruptible load management," *IEEE Trans. Power Syst.*, vol. 19, no. 1, pp. 683-689, Feb. 2004.

[42] M. M. He, E. M. Reutzel, X. Jiang, R. H. Katz, S. R. Sanders, D. E. Culler, and K. Lutz, "An architecture for local energy generation, distri-

bution, and sharing," *Proceedings of the IEEE Energy 2030 Conference*, November 2008.

[43] A. Capone, M. Barros, H. Hrasnica, and S. Tompros, "A new architecture for reduction of energy consumption of home appliances," *Proceedings of the European conference TOWARDS eENVIRONMENT*, March 2009.

[44] N. Gershenfeld, S. Samouhos, B. Nordman, "Intelligent infrastructure for energy efficiency," *Science, AAAS*, Vol. 327, pp. 1086-1088, 26 February 2010.

[45] M. A. A. Pedrasa, T. D. Spooner and I. F. MacGill, "Scheduling of demand side resources using Binary Particle Swarm Optimization," *IEEE Trans. on Power Syst.*, vol. 24, no. 3, pp. 1173-1181, Aug. 2009.

[46] N. Ruiz, I. Cobelo and J. Oyarzabal, "A direct load control model for virtual power plant management," *IEEE Trans. on Power Syst.*, vol. 24, no. 2, pp. 959-966, May 2009.

[47] T. Hubert, S. Grijalva, "Realizing smart grid benefits requires energy optimizatin algorithms at residential level," in *Proc. of IEEE Innovative Smart Grid Technologies (ISGT)*, 2011, pp. 1-8.

[48] Amir-Hamed Mohesenian-Rad, Vincent W. S. Wong, J. Jatskevich, R. Schober and A. Leon-Garcia, "Autonomous demand-side management based on game-theoretic energy consumption scheduling for the future smart grid," *IEEE Trans. on Smart Grid*, vol. 1, no. 3, pp. 320-331, Dec. 2010.

[49] A. J. Conejo, J. M. Morales and L. Baringo, "Real-time demand response model," *IEEE Trans. on Smart Grid*, vol. 1, no. 3, pp. 236-242, Dec. 2010.

[50] H. Alami, G. R. Yousefi and M. P. Moghadam, "Demand response model considering EDRP and TOU programs," in *Proc. of IEEE Transmission and Distribution Conference and Exposition*, 2008, pp. 1-6.

[51] J. G. Roos and I. E. Lane, "Industrial power demand response analysis for one part real-time pricing," *IEEE Trans. on Power Syst.*, vol. 13, no. 1, pp. 159-164, Feb. 1998.

[52] G. Xiong, C. Chen, S. Kishore and A. Yener, "Smart (in-home) power scheduling for demand response on the smart grid," in *Proc. of Innovative Smart Grid Technologies (ISGT)*, 2011, pp. 1-7.

[53] A. Molina-Garcia, F. Bouffard and D. S. Kirschen "Decentralized demand-side contribution to primary frequency control," *IEEE Trans. on Power Syst.*, vol. 26, no. 1, pp. 411-419, Feb. 2011.

[54] Y. Wang, I. R. Pordanjani and W. Xu , "An event-driven demand response scheme for power system security enhancement," *IEEE Trans. on Smart Grid*, vol. 2, no. 1, pp. 23-29, Mar. 2011.

[55] M. McGranaghan, A. Didierjean and R. Russ , "Commercial building load modeling for demand response applications," in *Proc. of CIRED 20th Internatinal Conference and Exhibition on Electricity Distribution - Part 1*, 2009, pp. 1-4.

[56] M. McGranaghan, A. Didierjean and R. Russ , "Business case for a consumer portal," in *Proc. of CIRED 18th Internatinal Conference and Exhibition on Electricity Distribution*, 2005, pp. 1-5.

[57] W-H Edwin Liu, Kevin Liu and Dan Pearson , "Consumer centric smart grid," in *Proc. of IEEE Innovative Smart Grid Technologies (ISGT)*, 2011, pp. 1-6.

[58] H. Hindi, D. Greene and C. Laventall, "Coordinating regulation and demand response in electric power grids using multirate model predictive control" in *Proc. of IEEE Innovative Smart Grid Technologies (ISGT)*, 2011, pp. 1-8.

[59] M. H. Nehrir, P. S. Dolan, V. Gerez and W. J. Jameson, "Development and validation of physically based computer mmodel for predicting winter electric heating loads," *IEEE Trans. on Power Syst.*, vol. 10, no. 1, pp. 266-272, Feb. 1995.

[60] N. E. Ryan, S. D. Braithwait, J. T. Powers and B. A. Smith, "Generalising direct load control program analysis: Implementation of the duty cycle approach," *IEEE Trans. on Power Syst.*, vol. 4, no. 1, pp. 293-299, Feb. 1989.

[61] R. E. Mortensen and K. P. Haggerty, "A stochastic computer model for heating and cooling loads," *IEEE Trans. on Power Syst.*, vol. 3, no. 3, pp. 1213-1219, Aug. 1988.

[62] R. E. Hatziargyriou, T. S. Karakatsanis and M. Papadopoulos, "Probabilistic calculations of aggregate storage heating loads," *IEEE Trans. on Power Delivery*, vol. 5, no. 3, pp. 1520-1526, July 1990.

[63] A. Molina-Garcia, M. Kessler, J. A. Fuentes and E. Gomez-Lazaro, "Probabilistic characterization of thermostatically controlled loads to model the impact of demand response programs," *IEEE Trans. on Power Syst.*, vol. 26, no. 1, pp. 241-251, Feb. 2011.

[64] N. Lu and D. P. Chassin, "A state-queueing model of thermostatically controlled appliances," *IEEE Trans. on Power Syst.*, vol. 19, no. 3, pp. 1666-1673, Aug. 2004.

[65] H. Saele and O. S. Grande , "Demand Response from household customers: Experience from a pilot study in Norway," *IEEE Trans. on Smart Grid*, vol. 2, no. 1, pp. 102-109, Mar. 2011.

[66] M. Hatori, "Peak-Shift Methods for Notebook Computers", *Proceedings of the International Symposium on Electronics and the Environment*, pp. 117-121, May 2004.

[67] T. Rausch, P. Palensky, "PROFESY: intelligent global energy management," *Proceedings of the IEEE International Conference on Intelligent Engineering Systems (INES 05)*, pp. 59-64, September 2005.

[68] K. Wacks, "Energy Management Rediscovered," *iHomes & Buildings, CABA*, pp. 10-12, Spring 2007.

[69] F. Kupzog, C Roesener, "A closer look on load management," *Proceedings of the 5th IEEE International Conference on Industrial Informatics*, pp. 1151-1156, November 2007.

[70] M. LeMay, R. Nelli, G. Gross, and C. A. Gunter, "An integrated architecture for demand response communications and control," *Proceedings of*

the 41st Annual IEEE Hawaii International Conference on System Sciences (HICSS '08), Waikola, Hawaii, January 2008.

[71] K. Wacks, "Home Area Networks for Electricity Demand Management," *iHomes & Buildings, CABA*, pp. 15-17, Summer 2008.

[72] A. Molderink, V. Bakker, M.G.C. Bosman, J. L. Hurink, G.J.M. Smit,"A three-step methodology to improve domestic energy efficiency," *Proceedings of Innovative Smart Grid Technologies (ISGT)*, January 2010.

[73] A Misra, H Schulzrinne, "Policy-Driven Distributed and Collaborative Demand Response in Multi-Domain Commercial Buildings," *Proceedings of the 1st International Conference on Energy-Efficient Computing and Networking, Passau, Germany*, April 2010.

[74] T. J. Lui, W. Stirling, H. O. Marcy, "Get Smart," *Power and Energy Magazine, IEEE*, Volume 8, Issue 3, pp. 66-78, May 2010.

[75] J. Xiao, J. Y. Chung, J. Li, R. Boutaba, J W. Hong, "Near Optimal Demand-Side Energy Management Under Real-time Demand-Response Pricing," *POSTECH University, Pohang, South Korea*, 2010.

[76] A-H. Mohsenian-Rad, V.W.S. Wong, J. Jatskevich, R. Schober, "Optimal and Autonomous Incentive-based Energy Consumption Scheduling Algorithm for Smart Grid," *Proceedings of the IEEE Conference on Innovative Smart Grid Technologies*, January 2010.

[77] J. Howard, H. Ham, N. F. Maxemchuk, "Smart Air Conditioners," *Proceedings of the Global Smart Grid Symposium, US-Korea Conference on Science, Technology, and Entrepreneurship (UKC)*, August 2010.

[78] S. Deering, M. Newborough, S. D. Probert, "Rescheduling electricity demands in domestic buildings," *Applied Energy, Elsevier*, Volume 44, Issue 1, pp. 1-62, 1993

[79] J. Abaravicius, "Load management in residential buildings: considering techno-economical and environmental aspects,"", *PhD thesis, Division of Energy Economics and Planning, Department of Heat and Power Engineering, Lund University*, 2004.

[80] R. Shawa, M. Attreea, T. Jacksonb, M. Kay, "The value of reducing distribution losses by domestic load-shifting: a network perspective," *Energy Policy, Elsevier*, Volume 37, Issue 8, pp. 3159-3167, August 2009.

[81] V. S. K. Murthy Balijepalli, R. P. Gupta, and S. A. Khaparde, "Towards Indian Smart Grids," in *IEEE TENCON Conference,* Singapore, Nov. 2009.

[82] PA Consulting Group. *The smart grid vision for India's power sector*. PA Government Services, Inc., pp. 69-83, July 2010.

[83] F. Rahimi, A. Ipakchi, "Demand Response as a Market Resource Under the Smart Grid Paradigm," *IEEE Trans. on Smart Grid*, vol. 1, no. 1, pp. 82-88, June 2010.

[84] F. C. Schweppe,"Spot pricing of electricity," Kluwer Academic Publishers, 1988

[85] V. S. K. Murthy, V. Pradhan, S. A. Khaparde, "Review of demand response under smart grid paradigm," *IEEE PES International Conference on Innovative Smart Grid Technologies*, Kerala, December 2011.[Under review]

[86] M. A. A. Pedrasa, T. D. Spooner and I. F. MacGill, "Scheduling of demand side resources using Binary Particle Swarm Optimization," *IEEE Trans. on Power Syst.*, vol. 24, no. 3, pp. 1173-1181, Aug. 2009.

[87] N. Ruiz, I. Cobelo and J. Oyarzabal, "A direct load control model for virtual power plant management," *IEEE Trans. on Power Syst.*, vol. 24, no. 2, pp. 959-966, May 2009.

[88] A. Capone, M. Barros, H. Hrasnica, and S. Tompros, "A new architecture for reduction of energy consumption of home appliances," *Proceedings of the European conference TOWARDS eENVIRONMENT*, March 2009.

[89] W. Zhang, and A. Feliachi, "Residential load control through real-time price signals," *Proceedings of the 35th Southeastern Symposium on System Theory* , August 2003.

[90] S. Tiptipakorn, and W. J. Lee, "A residential consumer centered load control strategy in real-time electricity pricing environment," *Proceedings of 39th North American Symposium*, October 2007.

[91] S. Tiptipakorn, W. J. Lee, and C. Mao, "Price Naming on Home Appliance Load Controls in Real-Time Pricing Environment," *Proceedings of Power and Energy Society General Meeting*, July 2010.

[92] K. P. Schneider, J. C. Fuller, and D. Chassin, "Analysis of distribution level residential demand response," *Proceedings of Power System Conferencce and Exposition*, March 2011.

[93] C. Roe, S. Meliopoulos, R. Entriken, and Sunil Chhaya, "Simulated demand response of a residential energy management system," *Proceedings of Energytech*, May 2011.

[94] K. P. Schneider, J. C. Fuller, and D. Chassin, "Analysis of residential demand response and double-auction markets," *Proceedings of Power and Energy Society General Meeting*, July 2011.

[95] D. O. Neill, M. Levorato, A. Goldsmith, and U. Mitra, "Residential demand response using reinforcement learning," *First IEEE International Conference on Smart Grid Communications*, October 2010.

[96] "Advanced Metering Infrastructure Overview and Plan", Rochester Gas & Electric Corporation, New York State Electric & Gas Company, Feb. 2007

[97] [Online]. Available: Demand Response: An introduction, http://www.sgiclearinghouse.org, 2006

[98] V S K Murthy Balijepalli, S. A. Khaparde, and C. Dobariya, "Deployment of DR in India," IEEE PES General Meeting, Minneapolis, USA, July, 2010.

[99] A. P. Agalgaonkar, C. V. Dobariya, M. G. Kanabar, S. V. Kulkarni, and S. A. Khaparde, "Optimal sizing of distributed generators in DR," in *Proc. IEEE Power India Conference,* Delhi, India, Apr. 2006.

[100] [Online]. Available: http://www.cdmindia.nic.in/ Nov. 2009.

[101] [Online]. Available: Official website of Council of Power Utilities in India, http://www.indiapower.org/images/finalpaper.pdf. Oct. 2009.

[102] [Online]. Available: http://mnes.nic.in/pdf/MNRE_REC_Report.pdf. Nov. 2009.

[103] [Online]. Available: www.mnre.gov.in/. Nov. 2009.

[104] [Online]. Available: Official website of Power Ministry of India, http://www.powermin.nic.in/. Aug. 2009.

[105] [Online]. Available: www.nwcouncil.org/energy/powerplan/6/final/*SixthPowerPlan*$_C$*h*5.pdf

[106] C. A. H. Aramburo, T. C. Green, and N. Mugniot, "Fuel consumption minimization of a microgrid," *IEEE Trans. Industrial Applications,* vol. 41, no. 3, pp. 673-681, May 2005.

[107] A. Steynberg (Eds.), *A Textbook on Fischer-tropsch Technology*. Elsevier, 2004.

[108] "Clean Alternative Fuels: Fischer-Tropsch", United States Environmental Protection Agency, EPA420-F-00-036, March 2002.

[109] L. Costa, G. Kariniotakis, A. Kamarinopoulos, and N. Hartziargyriou, "Renewable for Isolated Systems-Energy Supply (RES) and Waste Water Treatment", Report on *State of the Art on Low-Cost Innovative RES Technologies*, ARMINES and ICCS/NTUA, RISE Partners Jul. 2005.

[110] [Online]. Available: http://www.electratherm.com/ Nov. 2009.

[111] [Online]. Available: http://www.benchmarkenergy.in/Products.aspx Nov. 2009.

[112] *IEEE Draft Guide for Design, Operation, and Integration of Distributed Resource Island Systems with Electric Power Systems,* IEEE Standard P1547 Draft 4.

[113] *IEEE Standard for Interconnecting Distributed Resources to Electric Power Systems,* IEEE Standard 1547-2003.

[114] [Online] OpenADR Alliance, http://www.openadr.org/, Nov. 2012.

[115] [Online] M.A. Piette, S. Kiliccote, G. Ghatikar, "Linking Continuous Energy Management and Open Automated Demand Response," http://openadr.lbl.gov/pdf/1361e.pdf, Nov. 2008

[116] [Online] IEC PC118, http://www.iec.ch/dyn/www/f?p=103:7:0::::FSP_ORG_ID:8701, Nov. 2012.

In: Smart Microgrids
Editor: Hussein M. Khodr

ISBN: 978-1-62808-975-2

Chapter 4

APPLICATION OF THE GENETIC ALGORITHMS TO DESIGN THE PID CONTROLLER FOR A DC MOTOR SPEED REGULATION

Nedim Tutkun**[*] **and Dinçer Maden
Department of Electrical and Electronic Engineering,
Düzce University

ABSTRACT

Precise control of DC motors can only be achieved by a well-designed controller such as a closed-loop PID or similar control techniques. The PID controller is mainly employed to control DC motors driving industrial machinery but, its performance highly depends on the gain coefficients. Although the Z-N approach is one of the well-known deterministic techniques to find these coefficients, it requires long procedure. Alternatively, the GA and similar methods are proposed by a limited number of researchers to design the PID controller by identifying gain coefficients according to the pre-defined conditions due to complexity of the Z-N application. In this study, the GA method was used to build the PID controller for the speed control of a separately-excited DC motor. The results were compared with those obtained from employing the widely-used controller design techniques of Z-N and the

[*] nedimtutkun@duzce.edu.tr

PSO for different types of motors. The results have shown that the GA considerably improve the performance parameters of steady-state error, the rising time, the settling time and overshoot minimization with respect to other methods for a DC motor under consideration and it can easily be applicable to similar DC motors for the same purpose.

Keywords: DC motor speed control, PID controller design, Genetic algorithms, Particle swarm optimization, Ziegler-Nichols method

ABBREVIATIONS

GA	Genetic Algorithms
PID	Proportional Integral Derivative
Z-N	Ziegler-Nichols
PID	Proportional Integral Derivative
K_p	Proportional gain coefficient
K_i	Integral gain coefficient
K_d	Derivative gain coefficient
R_a	Armature resistance
L_a	Armature inductance
R_f	Field resistance
L_f	Field inductance
i_a	Armature current
V_s	DC source voltage
V_f	DC source voltage for field
e_a	Induced voltage in the armature windings or the back electromotive force
T_d	Developed torque
k_e	Induced voltage constant
k_t	Developed torque constant
ω_m	Motor's angular speed
θ_m	Motor's angular position
J_m	Motor's inertia torque
B_m	Viscous friction coefficient
N	Number of individuals ingenetic population.

1. INTRODUCTION

Today DC motors have been using in wide-range applications due to their simple control characteristics. In most applications, the PID control technique is largely in use to meet a number of design specifications although there are a variety of control techniques [1]-[4].The performance of the PID controller mainly depends on the gain coefficients of K_p, K_d, and K_i and tuning these parameters produce highly efficient control operation. There are few PID tuning methods such as the Z-N and the Cohen-Coon etc. which are conventionally used to determine these gain coefficients [5]-[6]. The Z-N technique is well-known among them and requires no system modelling for open loop and closed loop control systems. However, tuning gain parameters with this method is not straight-forward and must accomplish a certain procedure. Therefore, to carry out this procedure we have to find dead time, maximum slope, steady-state oscillation gain and period time etc. A limited number of researchers proposes alternative methods to design the PID controller for a variety of control systems from various aspects. These methods usually compare with the Z-N method in terms of design specifications such as small rising and settling times and less overshoot as well as their applicability and simplicity.

In this study, rather than using the Z-N method, the GA approach based on the stochastic optimization is applied for identification of these parameters in speed or position control of a separately-excited DC motor as shown in Fig. 1 [7]-[8].

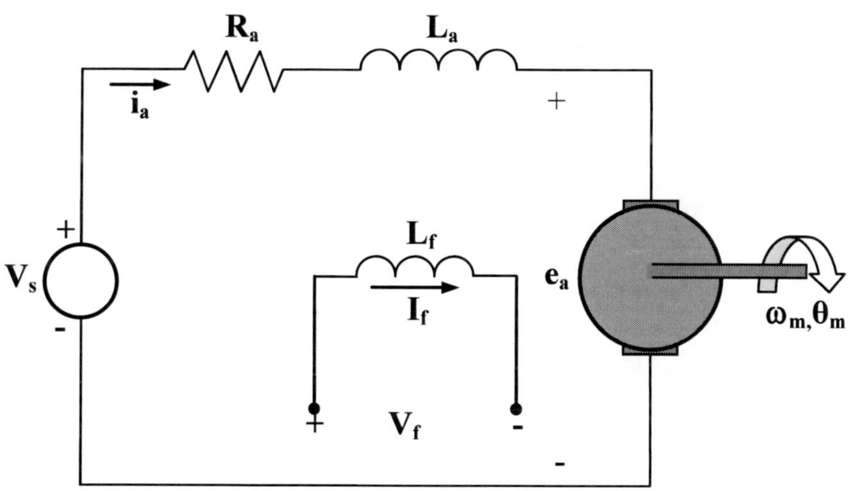

Figure 1. An approximate equivalent circuit for a separately-excited DC motor.

The necessary equations for modelling can be defined as

$$V_s = R_a i_a + L_a \frac{di_a}{dt} + e_a \tag{1}$$

$$T_d = k_t i_a \tag{2}$$

$$T_d = J_m \frac{d\omega}{dt} + B_m \omega \tag{3}$$

$$e_a = k_e \omega \tag{4}$$

$$\omega = \frac{d\theta}{dt} \tag{5}$$

where $k_t = k_e = K$.

Using the simple control system block diagram shown in Fig. 2, the open-loop response of the motor can be expressed as

$$\frac{\theta(s)}{U(s)} = \frac{K}{J_m L_a s^3 + (R_a J_m + B_m L_a)\, s^2 + (K^2 + R_a B_m)\, s} \tag{6}$$

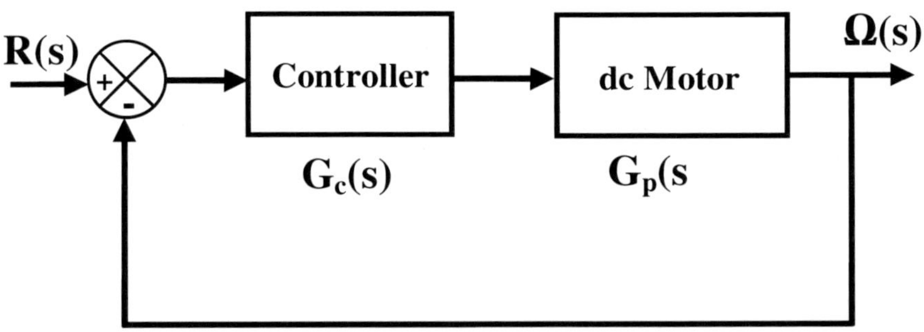

Figure 2. Block diagram of speed control of the separately-excited DC motor.

2. Estimation of the PID Parameters with the GA Method

The GA method is based on the natural evolution and selection, and works with a randomly generated population composed of the binary or real numbers. A new population is formed by selecting individuals among members of the initial population according to their fitness values through a selection criterion. Once the selection is completed, the crossover operation is implemented on the selected population to generate dissimilar individuals which may be possible solutions to the problem under consideration. Next, the mutation operation is applied to the current population to avoid early convergence. Finally, elitist strategy is preferably used to uphold the best individual for the next generation. Thus, the first generation is completed and the whole process is repeated until the termination criterion is met as illustrated in Figure 3.The number of generation can be increased and the advanced operators may be used if the best individual produces no satisfactory solution. In this study, the number of population individuals is 100, selection criterion is the roulette-wheel, crossover and mutation probabilities are 0.8 and 0.05 and their types are uniform and perturbation respectively. Moreover, the elitist strategy is applied for a single individual to improve the genetic performance by copying the best individual into next generation. The search space ranges from 0 to 1000 for each parameter and the fitness function to calculate the kth fitness value can be expressed as in the PID controller.

$$f_k = \frac{1}{\left\| \Omega_k \right|_{max} - R \right| + \varepsilon}, \ k=1, 2,\ldots, N \tag{7}$$

where $\left|\Omega_k\right|_{max}$ is the maximum of angular speed for k^{th} individual, R is the unity reference input signal, ε is the error margin to avoid making denominator zero and its value is10^{-5}.The GA software encoded by the MATLAB package was performed to find the optimal gain coefficients aiming to minimise the rise and settling times, and less overshoot for the separately-excited motor. It was designed to stop the looping process in a certain number of generation rather than using different termination criteria and it was repeatedly run until the desired solutions were obtained.

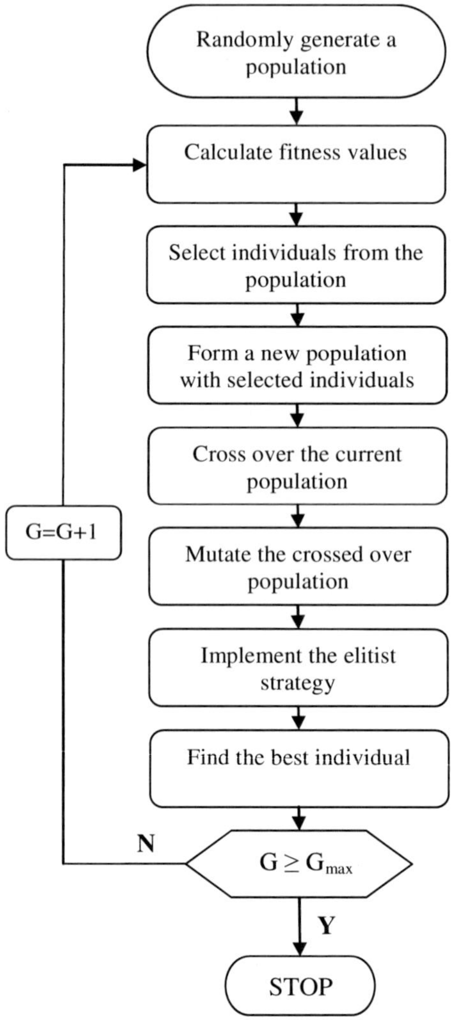

Figure 3. Flowchart of the genetic process.

3. Simulation Results

The results indicated that the determined PID parameters offer better control performance with respect to those obtained by the Z-N method for the 1st motor's parameters. I n the case of the DC motor used, the gain coefficients of k_p, k_i and k_d were found to be 86.4125, 73.6070 and 89.0518 respectively.

For a typical comparison, the Z-N open-loop tuning method was used to design the PID controller by determining the gain parameters. The GA results improved output response compared to the Z-N results both under and no load conditions. The uncontrolled and the Z-N output responses of the motor were obtained to verify the improvement on the PID controller as shown in Figs 4, 5 and 6. For the same motor parameters, maximum overshoot, the rise time and the settling time were estimated for both techniques and the obtained results are shown in Table 1.

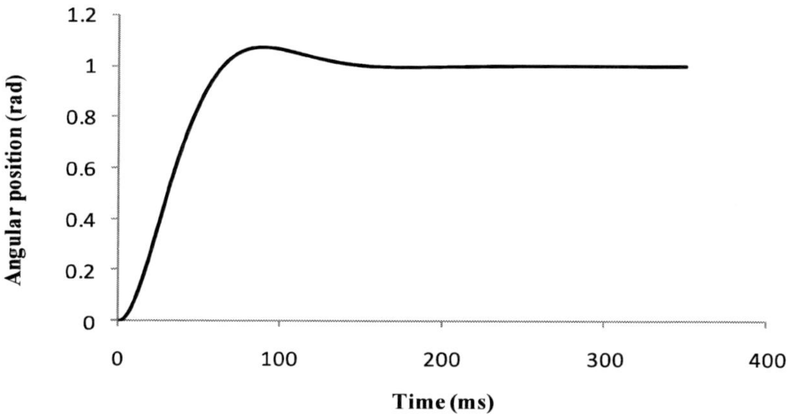

Figure 4. Uncontrolled unit step response of the separately-excited DC motor.

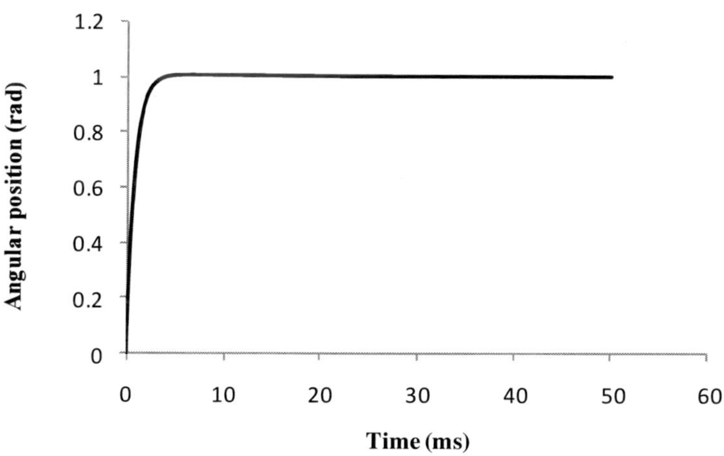

Figure 5. Unit step response of the separately-excited DC motor using the Z-N method.

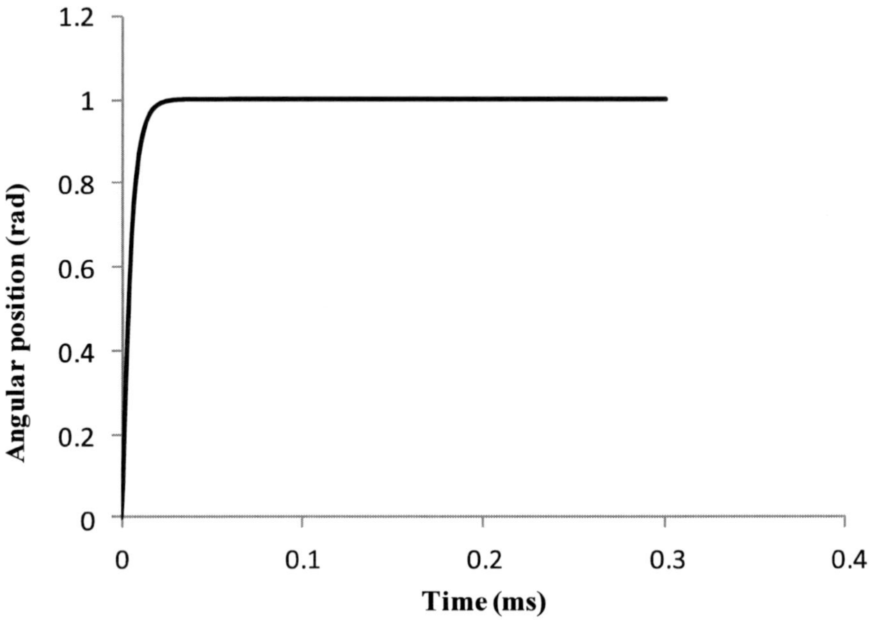

Figure 6. Unit step response of the separately-excited DC motor through the GA method.

Table 1. Comparison of the Z-N and the GA methods for given motor parameters [8]

Method	K_P	K_I	K_D	Overshoot (rad)	Settling time (ms)	Rise time (ms)
Genetic Algorithms	86.41	73.61	86.41	0.0009	0.049	0.036
Ziegler-Nichols	38.67	25.82	0.55	0.0065	10.550	1.830
Uncontrolled	1	0	0	0.0716	109.930	42.610

In order to obtain a fair comparison between the GA and the meta-heuristic methods, similar works based on the PID controller design of a DC motor are surveyed. It is seen from current investigation that the PSO is extensively used to tune PID gain coefficients among the meta-heuristic methods and a limited number of works based on the PSO algorithm is examined for a typical comparison. As a result of this, the current GA software was modified to obtain best PID parameters improving controller's performance such as less steady-state error, small settling and rise times, and low overshoot etc. The fitness function for the GA software was adapted by changing old motor parameters to new ones obtained from previous investigation in which the PID controller was designed by the PSO algorithm [9]. The GA results were obtained after a certain number of software run and they are illustrated in Table 2 for comparison. The GA results were compared PSO software encoded by the MATLAB package was performed to find the optimal gain coefficients aiming to minimise the rise and settling times, and less overshoot for the given DC motor. The GA results show better outcomes with respect to those obtained by the PSO. The controlled output response of the motor was shown in Figs 7. In addition, root-locus, bode diagram, Nichols chart and Nyquist diagram were also obtained to understand the DC motor system behaviour in Figs 8, 9, 10 and 11.

Table 2. Comparison of the PSO and the GA methods for given motor parameters [9]

Method	K_P	K_I	K_D	SSE (rad)	Settling time (ms)	Rise time (ms)
Genetic Algorithms	98.6056	94.3559	100.0000	0.00015	0.024	0.022
PSO	30.6575	12.6575	16.0979	0.0100	1.441	1.372
Uncontrolled	1	0	0	0.0213	356.00	79.00

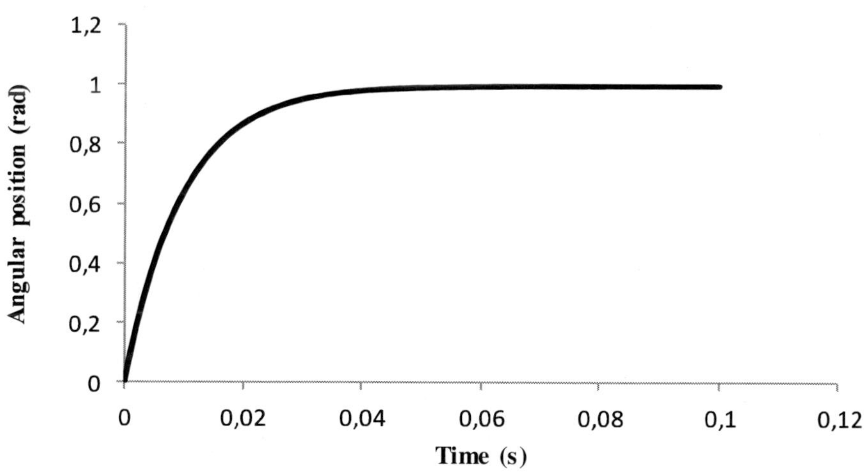

Figure 7. Unit step response of the DC motor under load through the GA method.

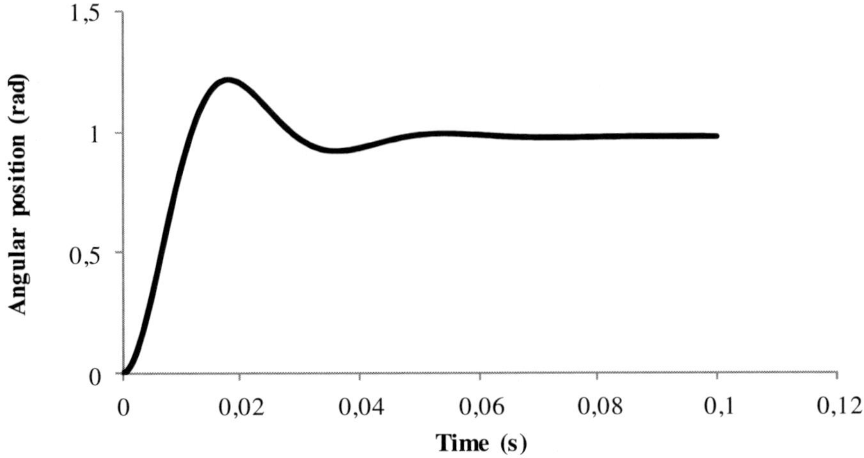

Figure 8. Unit step response of the DC motor under loadwithout a controller for 2nd motor's parameters.

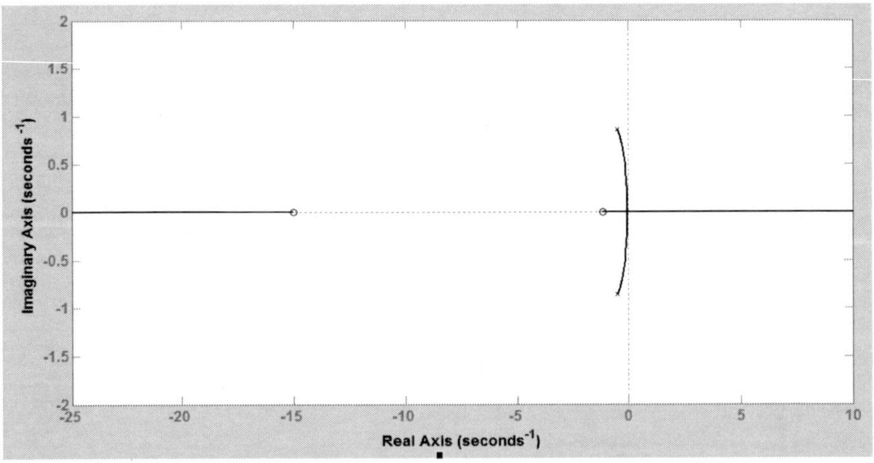

Figure 9. Root-locus of unit step response of the DC motor under load with the GA approach.

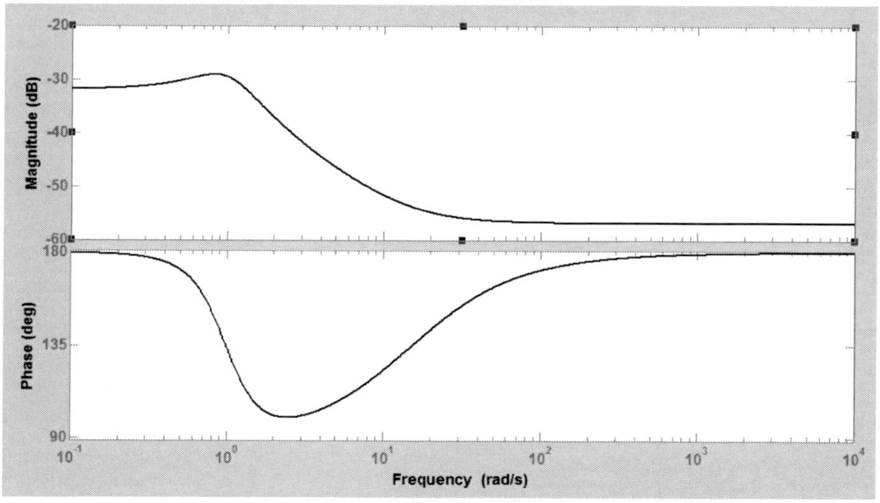

Figure 10. Bode diagram of unit step response of the DC motor under load through the GA method.

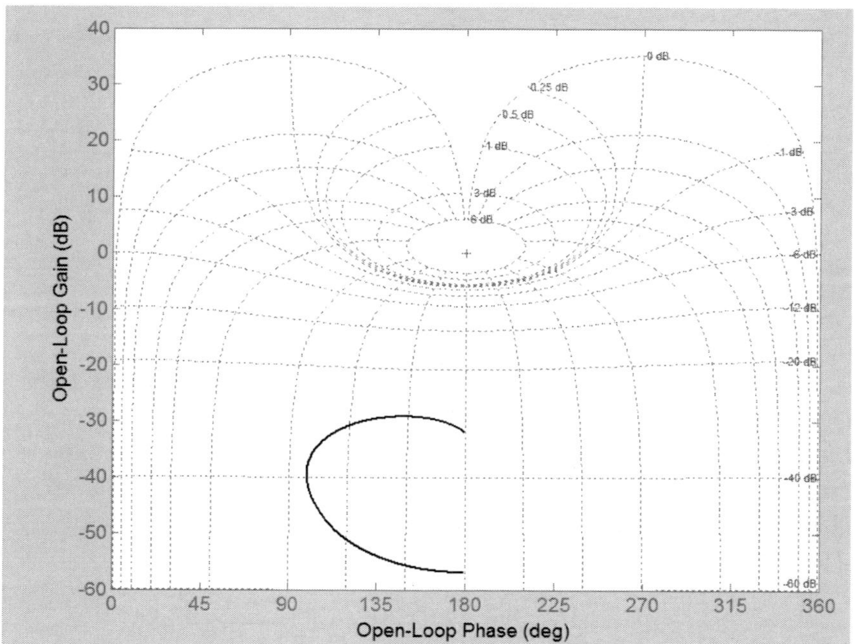

Figure 11. Nichols chart of unit step response of the DC motor under load through the GA method.

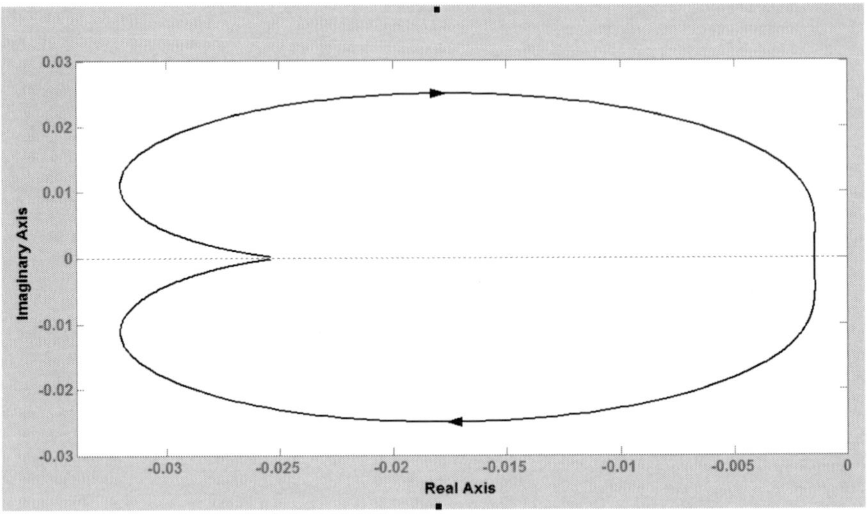

Figure 12. Nyquist diagram of the unit step response of the DC motor under load using the GA method.

The results also show that the GA method achieves to reduce maximum overshoot to almost zero, considerably minimizes the settling and rise time for the given motor's data with respect to those obtained by the Z-N method. Thus, the determination of the PID parameters with the GA improves the PID controller with respect to the Z-N method and this can be applied to a variety of similar problems.

CONCLUSION

The GA method can successfully be used to design the PID controller and the determination of the gain coefficients is very simple and more reliable than Z-N method and PSO technique in terms of minimising overshoot, settling time and rise time for the given motor's parameter. The results may also be applicable to the real-time systems such as DC motor driving industrial machinery for speed and position control.

REFERENCES

[1] K. Krishnakumar and D. E. Goldberg, "Control System Optimization Using Genetic Algorithms", *Journal of Guidance, Control and Dynamics*, vol.15, no.3, pp. 735-740, 1992.

[2] A. Herreros, E. Baeyens, J. R. Perán, "Design of PID-type controllers using multi-objective genetic algorithms", *ISA Transactions*, vol. 41, no. 4, pp. 457-472, 2002.

[3] P. Wang and D.P. Kwok, "Optimal design of PID process controllers based on genetic algorithms", *Control Engineering Practice*, vol. 2, no. 4, pp. 641-648, 1994

[4] N. Tutkun and D. Maden, "Determination of the PID controller parameters for a DC shunt motor using the genetic algorithm method", *National Conference on Electrical, Electronics and Computer Engineering (ELECO)*, pp. 330-333, 2010

[5] K. J. Åström, T. Hägglund, "Revisiting the Ziegler Nichols step response method for PID control", *Journal of Process Control*, vol. 14, no. 6, pp. 635-650, 2004.

[6] D. Valério, J. Costa, "Tuning of fractional PID controllers with Ziegler–Nichols-type rules", *Journal of Signal Processing*, vol. 86, no. 10, pp. 2771-2784, 2006.

[7] K. Ogata, *Modern Control Systems*, Prentice Hall,1987

[8] B. C. Kuo, *Automatic Control Systems*, Prentice Hall,1995

[9] Adel A. El-Gammal and Adel A. El-Samahy, *A Modified Design of PID Controller For DC Motor Drives Using the PSO, Power Engineering Conference,* Lisbon-Portugal, 18-20 March, pp. 419-424, 2009

APPENDIX

Motor's parameters:

R_a=4Ω, L=2.75μH, $J_m=3.2284\times10^{-6}$ kgm^2, $B_m=3.5077\times10^{-6}$ Nm/rad/s, k_e=0.0274 Vs/rad, k_t=0.0274 Nm/A [8].

R_a=7.56Ω, L=0.055H, J_m=0.068 kgm^2, B_m=0.03475 Nm/rad/s, k_b=3.475 Vs/rad, k_t=3.475 Nm/A [9].

In: Smart Microgrids
Editor: Hussein M. Khodr
ISBN: 978-1-62808-975-2

Chapter 5

PROTECTION AND AUTOMATION TRENDS IN MV AND LV SMART GRIDS

Nabil El Halabi[1], Samuel Borroy[1], Miguel García-Gracia[1] and Hussein M. Khodr[2]
[1]CIRCE, Research Centre for Energy Resources and Consumption, Zaragoza, Spain
[2]Qassim University, Engineering College, Department of Electrical Engineering, Buraidah, KSA

ABSTRACT

The operation of the electric power grid is expected to change over the coming years with the implementation of distributed generation and new technologies to improve efficiency and move towards smartgrids. In this chapter, some of the new trends to be implemented in smartgrids at medium and low voltage are discussed and requirements to address specific challenges for protection and automation are identified.

Advanced distribution automation and protection are introduced herein, along with their network architecture and arrangement of different types of IEDs proposed for smart grids. These schemes are intended to allow a reliable operation of distribution networks not only under grid-connected condition but also during autonomous operation.

Keywords: smart grid, MV & LV protection, fault location, restoration time, Advanced Distribution Automation

1. INTRODUCTION

One of the objectives of the automation of distribution networks is the optimization of system operation, minimizing losses and resolving potential scenarios of network overload. Likewise, the distribution network response under fault situations is an important issue for power quality, being the continuity of power supply a key factor for assessing the reliability of the network. The reliability indices, SAIFI (System Average Interruption Frequency Index) and SAIDI (System Average Interruption Duration) are both strongly influenced by the protection and automation process of detecting, locating and isolating fault, by following a proper sequence for service restoration.

The concept of network self-healing, aimed to be implemented in smart grids, involves the automation fault detection and isolation in order to reduce time and minimize disruptions number of operations necessary to isolate the faulted branch.

Figure 1 shows the evolution of SAIFI and SAIDI indices, due to unplanned outages, in some countries of the European Union in 2009. According to this data obtained from the CEER (Council of European Energy Regulators), the mean of the SAIDI in 2009 within the EU was approximately 200 min/year [1].

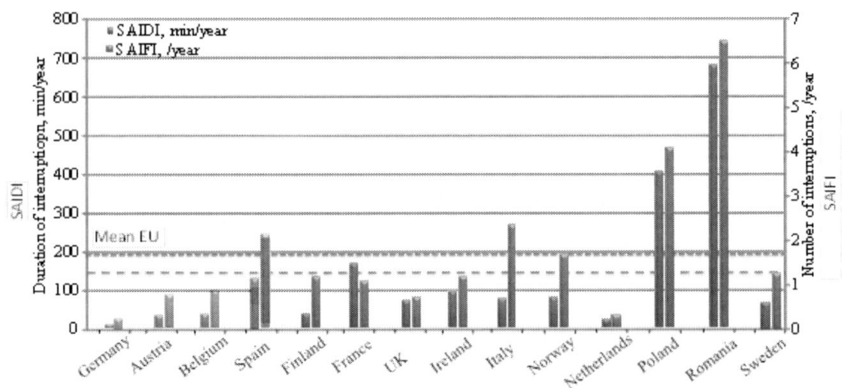

Figure 1. Quality indices of the European community energy as per CEER in 2009 [1].

The continuity of power supply is a matter of interest for any type of end user. For the industrial sector, power interruption of relatively short duration can lead to significant economic losses; while for home users may represent interruptions in heating, lighting and other everyday services [1], [2]. Currently, distribution networks lack or have an insufficient degree of automation. Moreover, their protection schemes are based on overcurrent functions which are subject to mal-operation if implemented in Distributed Generation (DG) scenarios.

Since the European electric power systems rapidly run up against their operational limits, the idea that a smart grid is not only needed but stands well within reach, has started being established as a major challenge for industry, scientists and policy makers around Europe. More specifically, the EU Smart Grids Research Agenda [3] sets among its priorities the transition from current power networks to smart grids that will be:

Flexible: fulfilling customer's needs whilst responding to the changes and challenges ahead.

Accessible: granting connection access to all network users, particularly for renewable power sources and high efficiency local generation with zero or low carbon emissions.

Reliable: assuring and improving security and quality of supply, consistent with the demands of the digital age with resilience to hazards and uncertainties.

Economic: providing best value through innovation, efficient energy management and "level playing field" competition and regulation, while stressing the need for distribution grids that are accessible to distributed generation and renewable energy sources, either self-dispatched or dispatched by local distribution system operators and facilitate dynamic control techniques and high levels of power security, quality, reliability and availability.

The smart distribution grid needs to evolve the current state of the power systems and include the integration of enhanced capabilities with a special focus on [4]:

Analytical software and advanced metering infrastructures, along with wide area monitoring and communication mesh networking infrastructures, using new system measurements for grid monitoring and forecasting activities.

Algorithms for the optimized mitigation of faults occurring in the distribution grid on the basis of genetic and heuristics algorithms for fast and accurate convergence.

Accurate model-based vulnerabilities profiling mechanisms for the diagnosis and cause attribution of the occurring faults both in the power and Information and Communications Technology (ICT) subsystems of the smart grid.

DG and Load forecasting on the basis of improved models to meet predicted generation with load and energy storage systems.

Numerous intelligent agents and communication devices integrated with power system monitoring and control applications, real-time data acquisition and smart meters and sensors will provide a significantly larger volume and new categories of data, such as wide-area phasor measurement information. This real-time data in joint with advanced data processing and visualization techniques will provide system operators a rapid grasp of the power system stability margins. Advanced Metering Infrastructure (AMI) [5] systems will provide an additional new source of relevant distribution status information, including loadings, voltage profiles, harmonics, and outage conditions, among other parameters.

Command and control platforms and optimization algorithms for distribution operations that will actually serve as hubs for many new self-healing features.

Advanced power electronic devices that will be employed on the level agents and communicate with central systems in order to adapt to real-time conditions and provide more secure and reliable control and protection of distribution grids and their constituent components.

High speed switching, throttling, modulating, and fault-limiting devices that will dynamically reconfigure the grid. This includes faster isolation and zoning, as well as rapid control of power flows in response to dynamic system challenges.

Islanding protection, with the main objective of detecting the islanding condition and isolating properly from the main grid. The tripping time for the islanding protection should be critically decided such that the two systems are successfully separated first before any out-of synchronism reconnection attempt.

Islanding protection is quite complicated since the Circuit Breaker (CB) that causes the loss of connection could be any breaker or isolator connecting the main source of supply to the DG site. The technological advancement has enabled the use of advanced low cost power electronics with integrated protection functions provided by

appropriate novel algorithms, including islanding configuration, stability, reliability and re-integration to the main grid.
Even though the islanded operation with DGs is usually not allowed since restoration by reclosing is difficult and power quality within the islanded section cannot be guaranteed, however, if the employed protection scheme is able to isolate the faulted segment and allow intentional power islands to operate with adequate protection, the reliability of the system can be significantly increased

To this end the introduction of ICT and smart-grid platforms supported by analytics, communication technologies and distributed intelligence, altogether formulating islanding protection operational management and automated control structures, stand as an ideal and significantly effective solution for the coordination and balancing of sources, storage and loads to produce a more reliable and secure decentralized power system in comparison with the traditional centralized approach.

2. Automation Scheme for Intelligent Distribution Network

The new trends in the field of electric power systems point to the development of intelligent distribution networks. Recent investigations outline that smart grids efficiency is approximately 60%, which supposes a 27% of improvement as compared with current distribution networks [6], [7]. Also, the development of intelligent distribution networks is a necessary condition for the implementation of distributed generation and electric vehicles. Figure 2.8 depicts the essential functions to be developed in smart grids.

In this context, the automation of the distribution network is oriented to the control and management of distributed generation, storage and manageable loads in low and medium voltage (LV & MV). The automation points to be considered in distribution networks are current transformers (CT), which are located in the isolation nodes of the current distribution networks, such as: border points (BP), first switching points (1P) and second switching points (2P). Figure 3 shows schematically these isolation nodes in a given MV network of length l [8].

Figure 2. Functions to be developed in smart grids.

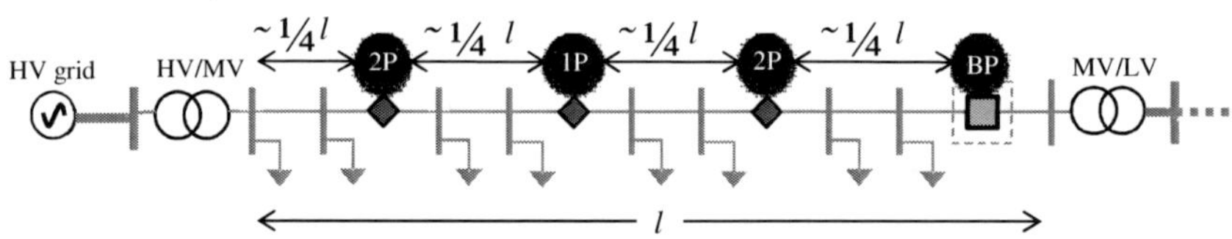

Figure 3. Border, first switching and second switching points in a MV grid of length *l*.

The operation procedure of distribution networks provided with these isolation nodes is as follows:

Border point (BP): It is located at the remote terminal of the distribution line which interconnects different lines of medium voltage (MV). It is normally open to maintain radial network operation, and closes in case any part of the circuit has been isolated from the source.

First switching point (1P): This point of measurement and isolation (CT) is located approximately at a length of the circuit that contains half of the installed load.

Second switching point (2P): These points are positioned such that they can isolate sections of the circuits containing quarter of installed load.

The automation scheme is aimed to enable an advanced monitoring provided by the aforementioned points as well as a self-healing function through a restoration algorithm. In this sense, load shedding is managed and ordered by algorithms that incorporate network reconfiguration and implemented at the main substation. Thus, the solution of automation in MV lines is:

Automatic operation of the first and second switching points (1P & 2P). The automatic action is performed by load isolation devices with an operation time below 100 ms.

Border points (BP) are also provided with voltage measurement as well which allows developing advanced algorithms for fault locations and for synchronization of distributed generation. Moreover, a continuous monitoring of voltage allows detecting reactive compensation in the network to enhance voltage profile.

At all points of isolation, including those off load isolators, are provided by directional faulted circuit indicators. These indicators are aimed to minimize the number of operations to detect the branch under fault.

3. ADVANCED DISTRIBUTION AUTOMATION FOR LV SMARTGRIDS

The automation of LV networks is focused on the installation of control devices aimed for monitoring and controlling the network operation. These devices are commonly known as IEDs ('Intelligent Electronic Devices') and are key concepts of intelligent networks with advanced distribution automation [7]. The network architecture and arrangement of different types of IEDs proposed for LV smartgrids is shown in Figure 4.

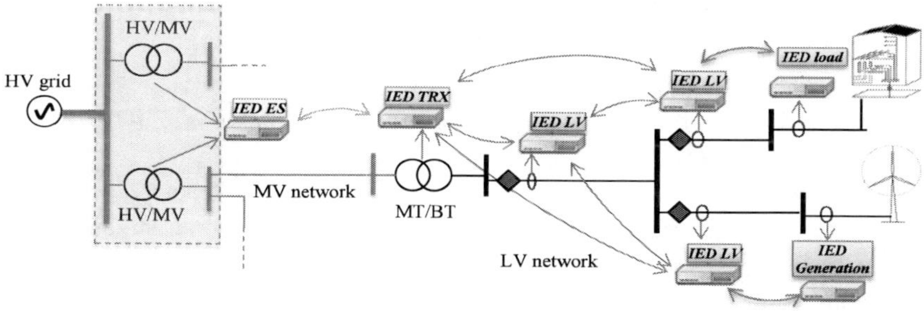

Figure 4. Network architecture and arrangement of different types of IEDs proposed for LV smartgrids.

IEDs implemented at the measurement points of the LV network (*IED LV*) are interconnected among themselves and with the IED of the transformer bus (*IED TRX*). In turn, these devices have a communication link with the main IED located at substations (*IED ES*), which supervises and controls all downstream IEDs.

As illustrates Figure 4, the IED ES manages the exchange of power between the MV network and the HV transmission system. Also, continuously evaluates the voltage profile of the MV network to determine the reactive compensation at each bus. This IED exchanges data with *IED TRX*, whose function is to optimize the power flow in the MV/LV transformers by balancing loads, micro-generation and micro-storage systems connected downstream of these transformers.

Moreover, the *IED LV* which monitors and reports the status of the LV branches to the *IED TRX* also executes the actions and setpoints received by the *IED TRX*. These IEDs deployed AMI functions for the acquisition and transmission of records made in the network. The AMI is addressed with the installation of sensors on the current transformers at each isolation points. These sensors allow obtaining the current measurement of each LV circuit and the status of the branch (connected/not connected) as well. The data recorded by sensors is transmitted to a receiver integrated in the *IEDs*, which in turn is connected to the upstream control devices of the network where the data is collected by the BT box sensors.

IEDs load/generation devices offer real-time monitoring and control of generation, storage systems and loads of the LV network, according to the orders of the IED that manages the LV branches where they are connected. The communication between the devices located in the transformer nodes and those at the point of loads and generation, is performed by a PLC ("*Programmable Logic Controller*") [9].

From orders placed by upstream devices, *IEDs load/generation* implement functions for local regulation. These functions are:

- Load management (passive or active) according to external signals of price, volume and consumption
- Managing generation and storage systems, by controlling reactive power and voltage, harmonic filtering, failures response, load shedding, power balance
- Historical record of consumption or generation
- Power measurement for load optimization
- Interface with other trading companies/users

Performance of IED Load/Generation

The *IED load/generation* are also known as smart nodes as they are devices that communicate to *IED ES* the measurement and state of the loads, generation, and storage, and are responsible of executing algorithms and procedures received by the control center. According to the philosophy of decentralized control in smart city projects, these smart nodes are able to act autonomously, but always in compliance with requirements that impose the control center before taking decisions. In any case, it is always allowed to the control center to take full control of a network branch monitored by the smart nodes.

4. Advanced Distribution Automation for MV Smart Grids

In MV smart grid, fault location and isolation are essential indicators to evaluate continuity of power supply and so, the performance of self-healing. This latter function refers to the capability to identify, isolate and restore the distribution network affected area with minimal or no manual intervention of operators. In essence, self-healing comes to represent the immune system of MV smart grids.

Firstly, it is argued that the self-healing function exerts a continuous real-time assessment to identify potential points of faults in the electrical system and initiate corrective actions immediately. However, the current state of distribution networks difficult the implementation of this self-healing function, mainly due to lack of cost-effective fault locators.

The new tendencies for implementing smart grid distribution point to the development of DMS ('Distribution System Manager') as shown in Figure 5.

The self-healing objective is to limit the disturbance impact to the smallest load area possible [6]. Consequently, self-healing can also enhance power quality indexes, taking such corrective steps as instantly transferring a customer from an instable source to a stable one. Fault location, voltage and power-quality problems, emerging dynamic instabilities, and other grid abnormalities will be quickly discovered and corrected. Advanced information models combined with new visualization tools will reveal congestion issues, overlays of failure probabilities, and resulting threat levels.

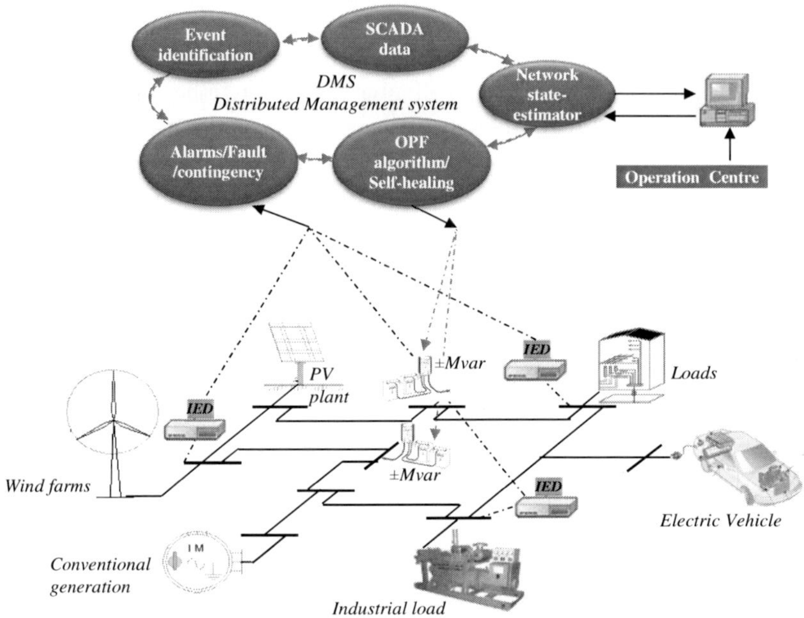

Figure 5. Implementation and operation of DMS MV smart grid.

In addition, the self-healing based-on grid will also have an enhanced ability to recognize high-risk transient conditions. When forecasted adverse weather and/or real-time contingency analyses are incorporated into a probabilistic model, grid operators will better understand the risks of each decision as action accordingly to minimize them.

Performance of IED ES and IED Trx

The *IED ES* and *IED Trx* implement shelf-healing algorithms and control of the distribution network. These IEDs include the following main functions for each of the technologies involved in the automation of the network:

AMI Functions

Data acquisition and supervision (Energy, V, f ...)
Anti-fraud or cyber-attack risk
Phase balance
Power limits

ADA: Advanced Distribution Automation

Self-healing: joint detection, isolation, power restoration
Optimal operation of the network by minimizing power losses
Disclaimers management
Online automation

DER: Distributed Energy Resources

Regulation or possible limitation of active power generation
Voltage regulation
Reactive compensation
Possible generation and load shedding
Wind, sun and load forecasts

Also, smart grid projects will implement gateway to transfer *IED load/generation* information to *IED ES* and control center for monitoring and diagnosis, through the same communication protocol.

5. Fault Management in Smart Grids

A high degree of DG penetration (more than 20%) as well as DER integration have had a considerable impact on the operation, control, protection and reliability of the existing distribution grid [10]-[12]. These issues must be thoroughly assessed and addressed before allowing DGs integration in to the electricity market. This is necessary for utilizing full DG potential for generation augmentation, for enhancing power quality and reliability and for providing auxiliary services such as active and reactive reserve, manageable loads and power supply restoration [13]. The task that is critically affected by DG penetration is protection coordination of the distribution grid. Conventional overcurrent protection schemes are designed for radial distribution systems with unidirectional current flow. However, the connection of DG in to distribution networks convert the singly-fed radial networks to complicated ones with multiple sources, enabling bidirectional fault contribution.

Currently, available technical recommendations such as, G83/1 [14], G59/1 [15], IEEE Std.1547 [16] prescribe that DGs should be automatically disconnected from the MV and LV distribution networks in case of tripping of the Circuit Breaker (CB) supplying the feeder connected to the DG. This is known as the anti-islanding feature. This is incorporated as a mandatory feature in the inverter interfaces for DGs available in the market. As the DGs are not under direct Distribution System Operator (DSO) control, use of anti-islanding protection is justified by the operational requirements of the DSOs. However, with greater DG penetration, automatic disconnection of DGs for LoG (Loss of Generation) situation drastically reduces the expected benefit of DGs for:

1) maintaining power quality and reliability
2) enhancing system security and
3) providing several ancillary services. It also leads to unnecessary loss of DG power in the event of grid faults [16].

In hierarchal distributed control strategies, the lower level of control can be independently operated and disconnected from the higher level in order to allow numerous microgrids (often in an islanded operation) provided that they have capability to balance supply and demand locally with acceptable power quality and security guarantees, as per local system requirements. Such control-independency enables zones of the network to be operated in two operation modes, autonomous (islanding) or grid-connected. This flexibility increases the reliability of power supply since their internal electricity resources can be used to supply their own demand during main grid disturbances. Following fault inception, the restoration algorithm activates an autonomous operation of generation/load by creating two microgrids (microgrid 1 and microgrid 2) as shown in Figure 6.

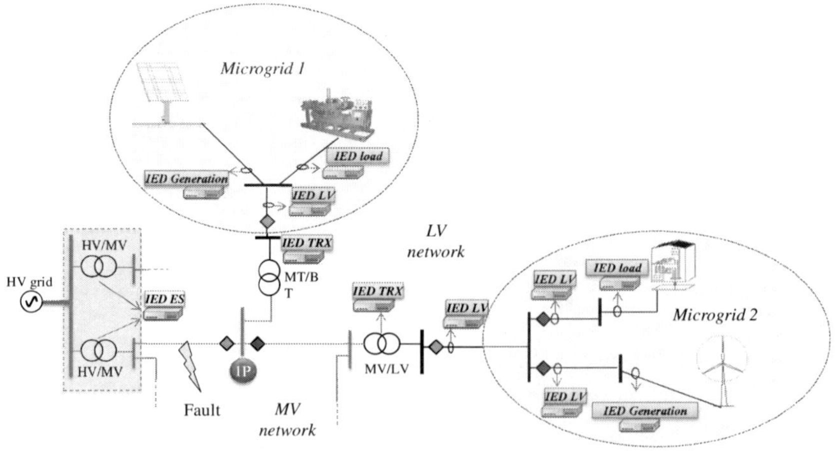

Figure 6. Autonomous (islanding) operation of MV and LV smart network.

Distributed intelligence and decentralized management and control could also add benefits to the rapid restoration of the grid in case of fault incidents. Restoration of loads that were initially disconnected can be relatively lengthy because restoration procedures rely on methods executed by controllers and are currently applied in small scale. For further enhancing security of supply in the distribution grid, adaptive and integrated protection/control systems are required to provide flexible, optimized response to network fault conditions, unusual transient behavior and restoration/reconfiguration for post-event recovery, incorporating temporal parameters regarding the duration of the fault and the need for further response.

Novel algorithms are required to predict and recognize eventual new phenomena brought about by complex interactions between large numbers of generator controllers, network management controllers (typically power electronics based) and active loads. It is important to highlight the need for the development of methodologies for system state monitoring based on wide area measurements so that this information can be used in conjunction with advanced active control and protection systems

The sequence for achieving autonomous operation is as follows:

1) Fault inception on the MV branch, between the CT and the first switching point, as depicted in Figure 5.
2) Fault detection algorithm implemented in *IED ES* and *IED Trx* enables the self–healing algorithm. Fault location algorithm allows the self-healing algorithm to determine initial constraints for forming microgrids.
3) Protection devices isolate the branch under fault by tripping 1P disconnector.
4) *IEDs load/generation* determines conditions for power balance and communicate measurement to the *IED Trx* which executes real time power flow to determine network constraints and availability of generation to form microgrids, and to maximize continuity of supply.
5) The results of the power and autonomous algorithm of *IED Trx* is forwarded to the *IED load/generation* for possible load or generation shedding, while the LV transformer CB is ordered to trip.
6) Once isolated and cleared the fault branch, the *IED ES* starts the sequence of normal operation restoration by interconnecting microgrids and, finally, recovering the normal operation on the MV network, while the *IED Trx* applied the same restoration into the LV network.

As it becomes apparent from the above analysis, an efficient formulation of microgrids as part of a self-healing smart grid process is a factor of the utmost importance towards the ultimate goal of fully recovering the grids normal operation. To achieve this implementation successfully, smart grid's self-healing should integrate the following attributes:

perform fast fault detection and precise cause attribution,
automatically determine the optimum dynamic and adaptive islanded micro-grids,
guarantee the completely functional operation of the islanded microgrid,
and re-integrate the dynamically formulated islands to the main grid with the utilization of automatic interconnection and synchronization power electronics and restoring its normal operation

These attributes require the design and application of new wide area monitoring and coordinated active grid control concepts that will ensure security and reliability of power supply through the adoption of dynamic and adaptive islanding protection management/control approaches. In addition, advanced fault detection mechanisms and prediction algorithms are needed, along with highly distributed and AMIs, for rapid restoration and self-healing of the grid.

CONCLUSION

Available technical recommendations such as, G83/1 [14], G59/1 [15], IEEE Std.1547 [16] prescribe that DGs should be automatically disconnected from the MV and LV distribution networks in case of tripping of the Circuit Breaker (CB) supplying the feeder connected to the DG. This is known as the anti-islanding feature and represents a drawback in advanced distribution automation techniques.

In MV & LV smart grids, distributed intelligence and decentralized management and control add benefits to the rapid restoration of the grid in case of fault incidents. The automation scheme is aimed to enable an advanced monitoring of measurements as well as a self-healing function through a restoration algorithm. Autonomous operation and self-healing features are key in attaining the full automation of electric networks.

In this sense, these concepts of automations are aimed to reduce restoration time and minimize disruptions number of operations necessary to isolate the faulted branch. Hence, power supply quality and reliability are enhanced.

REFERENCES

[1] CEER, Council of European Energy Regulators. *5th CEER Benchmarking Report on the Quality of Electricity Supply 2011*. Abril, 2011.

[2] Miguel García-Gracia, Nabil El Halabi, Adrián Alonso and M.Paz Comech (2011). Harmonic Distortion in Renewable Energy Systems: Capacitive Couplings, Power Quality Harmonics Analysis and Real Measurements Data, Gregorio Romero Rey and Luisa Martinez Muneta (Ed.), ISBN: 978-953-307-335-4, InTech, DOI: 10.5772/17677.

[3] European Technology Platform SmartGrids, Strategic Research Agenda: Update of the SmartGrids SRA 2007 for the needs bt the year 2035. March, 2012.

[4] Planas, E., A. Gil-De-Muro, J. Andreu, I. Kortabarria, and I. Martínez De Alegría. General Aspects, Hierarchical Controls and Droop Methods in Microgrids: A Review. *Renewable and Sustainable Energy Reviews*, 2013; 17: 147-159.

[5] Sreenadh, S.S., Depuru R., Wang, L., Devabhaktun, V. Smart meters for power grid: Challenges, issues, advantages and status. *Renewable and Sustainable Energy Reviews*, 2011; 15 (6): 2736-2742.

[6] Camacho, E, et al. Control for Renewable Energy and Smart Grids. The Impact of Control Technology, Samad T. and Annaswamy AM (eds.), *IEEE Control Systems Society*, 2011 [Online]: www.ieeecss.org.

[7] Electricity Advisory Committee-U.S. Department of Energy. Smart Grid Enabler of the New Energy Economy. 2008, [Online]: http://www.oe.energy.gov/DocumentsandMedia/final-smart-grid-report.pdf.

[8] El Halabi, N. *Fault locators for distribution networks, thesis dissertation*, University of Zaragoza, Zaragoza, 2012.

[9] Khodr HM, El Halabi N, García-Gracia M. Intelligent renewable microgrid scheduling controlled by a virtual power producer: A laboratory experience. *Renewable Energy*, 2012; 48, pp. 269-275.

[10] Lasseter R.H. and Paigi P. Microgrid: a conceptual solution. *Proc. of Power Electronics Specialists Conference (PESC)*, Aachen, Germany, 20- 25 June, 2004, Vol.6, pp. 4285- 4290

[11] Zhao, J-H., Foster, J., Dong, Z-Y., Wong, K-P. Flexible Transmission Network Planning Considering Distributed Generation Impacts. *IEEE Transactions on Power Systems*, 2011; 26 (3): 1434 – 1443.

[12] Khani, D., Sadeghi, A.Y., Madadi, H.K. Impacts of distributed generations on power system transient and voltage stability. *International Journal of Electrical Power & Energy Systems*, 2012; 43 (1): 488-500.

[13] Lilley, W., Hayward, J., Reedman, L. Chapter 7 – Realizing the Potential of Renewable and Distributed Generation, in Smart Grid: Integrating Renewable, Distributed & Efficient Energy, Sioshansi F.P. *Academic Press Elsevier*, Oxford, 2012: 161-183.

[14] Engineering recommendation G83/1, Recommendation for the connection of small-scale embedded generators, London, 2003.

[15] ER G59/1 Recommendations for the connection of embedded generating plant to the Regional Electricity Companies' distribution systems, London, 2003.

[16] IEEE 1547 Standard for Interconnecting Distributed Resources with Electric Power Systems, USA, 2003.

[17] Chowdhury, S.P., Chowdhury, S., Chui Fen Ten, Crossley, P.A.. Islanding protection of distribution systems with distributed generators — A comprehensive survey report. *Proccedings of the IEEE Power and Energy Society General Meeting*, Pittsburgh, 20-24 July, 2008, pp. 1-8.

In: Smart Microgrids
Editor: Hussein M. Khodr

ISBN: 978-1-62808-975-2

Chapter 6

OPTIMAL RECONFIGURATION OF ELECTRICAL DISTRIBUTION NETWORKS MAKING THE SCADA MORE INTELLIGENT IN THE CONTEXT OF THE SMART GRIDS

Hussein M. Khodr*
Qassim University, Engineering College,
Department of Electrical Engineering, Buraidah, KSA

ABSTRACT

Distribution networks are normally meshed in design but the operation is nearly always configured radially. Their configurations may be varied with manual or automatic switching operations for the loads supplied at the minimum cost of possible line losses, increasing system security and enhancing power quality. Reconfiguration consists in changing the network configuration by opening/closing sectionalizing - and tie-switches so that the networks become radial in operation. The objective minimizes power losses, balancing load among feeders and subject to constraints such as capacity limit of branches, minimum and maximum power limits of substations or distributed generators, minimum deviation of bus voltages and radial optimal operation of networks. The

* Email: khodr.hussein@gmail.com

problem is formulated as Mixed-Integer Linear programming problem and solved by LINGO. This model can be used into the SCADA platform as online application to solve the overloaded conditions of the networks or to isolate the fault, which may occurs, making the network more intelligent acting as a smart network controlled by the SCADA and obeying all the technical constraints. The methodology has been applied for a large real world power distribution system with a significant saving in active power losses and in term of operation of the network, demonstrating the effectiveness of the proposed methodology integrated with the SCADA system.

Keywords: Distribution system, Reconfiguration, SCADA, intelligent networks, Smart grid operation

1. INTRODUCTION

Operation planning methods in electrical distribution networks have evolved since they first appeared until this century, thanks to the development of calculation techniques and methodologies. In general terms, we can outline two different periods of evolution:

- Up until the 1960s, calculation tools were not able to solve complex mathematical problems, so the planner's experience played a major role and solutions which could satisfy technical requirements were sought.
- As a result of the arrival of computers, and their high ability to process complex numerical models, it was possible to perform technical and economic calculations. Initially, the use of computer devices was expensive and complicated to the planner.
- From the 1980s onwards, the arrival of microcomputers simplified man-machine interfaces and the use of computer science became more economical. Nowadays, it is possible to solve complex mathematical problems with fast results and an easier interaction with the computer machine, allowing us to introduce several variants aiming to obtain the best technical and economic solution.
- Within this context, several methods have been developed, and some of them have been supported by new calculation tools which supersede mathematical techniques previously used in operation and planning of electrical distribution networks (reconfiguration). Among

them, the most prominent are Natural Methods, such as neural networks, and genetic and evolutionary algorithms [1].

In this section, a historical study regarding the development of planning methods for electrical distribution networks and its critical analysis will be introduced. Planning methods first appeared in early 20th century with the development of electrical networks. In the early years of the 1960's, the planner's experience and talent played a key role. Calculation tools were limited to slide rules and mechanical calculators, which hampered the performance of complex mathematical calculations. However, at the end of this stage, new project methods for networks using computer algorithms, emerged to assist planners when operating networks [2].

From the 1960s onwards, there have been several publications, showing an intense work related to planning electrical distribution networks. These include different methods, such as:

- Evaluation methods [3-5]. Define as starting point the preparation of networks variants which comply with the technical requirements of quality and reliability of electric service. These variants are submitted to a cost evaluation in order to select the most economical.
- Methods which use power flow calculation results [6-8]. Authors base their decisions on yielded results from the calculation of the power flow of electric network. The method is based on the submission of an early composed network to an elimination process in which nonessential elements are removed, in order to keep the service running searching for the most economic network. Then, a radial network is sought from a network composed by all its elements closing several mashes. The process of removing nonessential elements consists of a loop process in which power flow analysis is applied. In each loop, lines carrying less power flow to each node are eliminated, thus ensuring that the network feeds all loads and complies with all technical restrictions such as voltage drop, transformers technical limits and lines limits. The process is finished when deleting an element does not meet the technical constraints, or makes the network not capable of feeding all loads.
- Methods using linear programming techniques [9-45]. The purpose of making a mathematical model representing the network that is to be planned is to minimize investment costs and minimize energy losses. These are a quadratic function of the power which circulates through

network elements. In order to use linear modelling, authors set out losses expressions in different ways [46]. All admit that the mathematical model is a mixed-integer type. Some authors [47-51] combine linear mathematical optimization techniques with heuristic criteria. Mathematical optimization enables us to make decisions based on the analysis performed by the planner. Others [52-58] argue that the most widely accepted technique is the Branch and Bound method.

- Methods which use graph theory [59, 60]. These methods are special cases of the models which use linear programming. They have the same optimization purpose, but represent the network according to an oriented graph, consisting of collections of nodes and edges. Graph theory is applied to solve the mathematical model.
- Methods using optimization techniques [61, 62, 63, 64, 65, 66, 67, 68]. They represent mathematical models that include nonlinear relationships of cost, capacity and the exact expression of Joule losses. Generally, they require specific techniques, according to the resulting model, and yield an approximate solution. Authors using linear programming claim that the inaccuracy of the approximate solutions of nonlinear methods, do not justify its application. Nevertheless, in [69, 70] dynamic programming is used to envision radial circuits and scaling the capacities of substations.

During the 1980s, and for the past few years, planning techniques for electric networks have been developed considerably. Advances in computer science played a key role in this step, improving already used techniques, and new emerged techniques. In some cases, methods are based on the same principles as previously developed methods, using the advantages of computer science. In other cases, they are based on new optimization techniques that have emerged in recent years. In the first cases, the use of heuristic evaluation techniques [71] and the use of economic-mathematical modelling with heuristic criteria stands out. These methods are referred to as mixed methods. In others, the use of natural methods [71, 72]: neural networks and genetic algorithms [73, 74, 75, 76, 77, 78, 79] are stressed.

Over the past few years, different methods to solve the problem of electric power distribution networks planning have been published in specialized literature. According to a recent classification, published in [81], methods could be divided into two categories: Electrical, if they use network features to organize the search for the best solution, and Natural if said search is determined by similar conditions to those existing in the natural world.

Electrical networks planning methods are also divided into three groups:

- Construction methods: These methods follow the strategy of reaching the most economic network starting from the nodes and their loads, and then adding elements which result in lead to the most economical solution until completing the minimum required to offer the desired service. There are different methodologies to this process which are in some way submitted to a technical-economic evaluation.
- Destruction methods: This method starts by defining an initial network which contains as much elements as possible. This sort of methods develop calculation procedures which reduce the network to their minimum, ensuring that there is a supply which meets the required quality and the lowest possible operating costs.
- Branch exchange or permutation methods: They start off from a network which meets the technical requirements, and in which there is a process of re-arrangement or permutations of its elements, giving rise to the change of power flow paths, evaluating the costs until the most economical arrangement has been found. These methods are used more in the reconfiguration of networks than on its planning.

Natural methods mentioned in the specialized literature are essentially associated with the neural networks [82, 83] and genetic algorithms [84]. However, other meta-heuristics techniques have also appeared, such as *Simulated Annealing* [85, 86, 87], *Branch Exchange* [88, 89, 90, 91], *Ant-colony System* [92, 93], *Evolutionary Computation* [94, 95], *Fuzzy Logic* [96, 97] and *Tabu Search* [98-100].

This following chapter is organized as follows:

It starts off in Section 1 (Introduction), consisting of an introduction to the topic of optimal reconfiguration of electric power distribution networks. It gives an overview of the evolution of methods used in the reconfiguration of electric power distribution networks and the need for its optimization will be justified.

Next, in Section 2 (Reconfiguration) reconfiguration of electric power distribution networks (REPDN) will be discussed. The notion of optimization, incorporated into the use of REPDN, will be introduced, and the technical limits of its basic equipment (lines, transformers, circuit breakers, isolators, etc.) will be referred. There will also be a brief approach to some additional aspects associated with the development of a project of this type of networks.

Section 3 (REPDN problem formulation) covers the central aspects of this chapter, describing this topic in more detail than in previous sections.

Initially (section 3.1 – Objective function), a mathematical expression of the objective function will be presented; the purpose is to minimize its variables and coefficients, and linearize the function of losses which is a quadratic function of the current. This linearity is obtained by including the current's economic density multiplied by the conductor section.

In section 3.2 (Constraints) the question of REPDN operation will be presented in detail, defining the technical limits to each network element according to standards established by regulatory authorities, and local operating distribution companies. These limits correspond essentially to the mathematical formulation of technical constraints, either in lines or in transformer stations and/or substations, thus, establishing a set of constraints to the REPDN operation, and therefore yielding a complex problem which ends in an optimal reconfiguration of the distribution system implemented into the SCADA system.

After formulating the problem, it is necessary to identify an optimization technique in order to solve it. To achieve this, a software optimization called *Optimization Modelling Software for Linear, Nonlinear, and Integer Programming* (LINGO) is used. Next, the physical meaning or interpretation of the results are presented and discussed.

In addition to solutions obtained from the network which is under study in this chapter, a real study case of a grid reconfiguration is analysed. This case highlights the current and subsequent situation following its reconfiguration, concerning its electric network parameters such as power, energy and voltage levels losses, corresponding to a new topology network.

Next, there will be a reference to aspects which will affect the evolution of the distribution network reconfiguration, highlighting the tendency to the increasing efficiency of entities which manage electric energy distribution networks, higher customer satisfaction, and improvements on the quality of service and increasing concerns/needs of sustainably in terms of development/growth. Furthermore, developments in other areas, which will determine the ability of optimization methods to solve these problems, including the processing capacity of computers to deal with large volumes of information and the introduction of smarter meta-heuristics techniques.

To finish this chapter, there will be a final synthesis (section 4 Conclusions). Starting from the presentation of a situation-problem, in this specific case, the reconfiguration of an existing electric network, comparing the results of the power flow before and after its reconfiguration, which allows the engineer, student, planner to decide which solution is the best. Thus, we use prior knowledge so that we can integrate it within the students' or engineers understanding network, that is, practical knowledge. They are the ones responsible for taking their own conclusions; build their own knowledge regarding approached themes, always taking explanations (expository teaching method) and oriented questions asked by the teacher throughout the class (teaching based upon discovery) into consideration.

2. Reconfiguration of Electric Power Distribution Networks

In order to respond to an increasingly demanding market, concessionary companies of electric power distribution networks have been continually trying to improve the quality of their product, that is, of the energy they provide to their users. The growing automation of distribution systems presents an opportunity to change the configuration of a network more easily by manoeuvring shut-off and cross section devices, guaranteeing actions which allow to operate the system in the best suited way in different situations, which leads to a decrease in losses, an improvement of the use levels in network elements, thus guaranteeing suitable tension levels. Considering that power supply cut offs are inevitable, whether when executing expansion works, during preventive maintenance actions in network components or because of the action of a protection device due to a defect, there should be a manoeuvring plan for the reconfiguration of the system, so that the area affected by the electrical power failure is minimized. The goal is to restore energy supplies to users who are down in this area as soon as possible by manoeuvring shut-off and cross section devices in the network, keeping the radial configuration of the system. In the upstream loads of the affected area, the supply of electric power will be restored through the action of protections and the reconfiguration of the system will not probably be necessary.

Nowadays, thanks to the opening of the electric sector/re-estructuration and its new divided structure into companies which sell electric power and companies which provide services in order to allow its distribution, there is a need to develop competitiveness between companies which produce electric power, those which transport it and, especially, those which are in charge of the distribution of electric power. Thus, it is necessary to know the behaviour of loads and electric parameters on a daily basis.

In order to reach these goals, it is necessary to know the state of the network for all the operation's time intervals. Thus, power suppliers may apply different fees, mainly driven by the price as a tool to the management of electric load, which means, it penalizes consumers who use energy during peak energy times because they are contributing to overload the network and, thereby, allowing the entrance of inefficient power generation stations into the system. Within the open market, suppliers must optimize electric power distribution networks, minimizing their operation costs in order to assure a suitable service with high-quality levels.

Distribution networks are usually designed as loops, but the operation is always radial. Their settings may be varied by commutation, hand or automatically operated in order to supply loads at the lowest possible cost for line losses, increasing the security of the system and the quality of the service [101], including the busbars voltage levels. This reconfiguration is based upon the change on the line's configuration by opening/closing sectionalizing and tie-switches so that networks can become radial in operation.

Several researches have been carried out in order to minimize losses in the reconfiguration area of electric power distribution networks [102-107], which have optimized voltage profiles thanks to a suitable operation of these networks.

In order to reduce losses, as suggested by Merlin and Back [103], it was necessary to reconfigure the distribution network. In order to achieve this, a Branch and Bound optimization technique was used so that the radial configuration could be determined with as little active losses as possible. In this method, all switches are initially closed in order to form a meshed network, and are then successively opened to restore a radial configuration through a destructive method.

Shirmohammadi and Hong [104] have been suggesting a heuristic algorithm based on Merlin and Back's algorithm [103]. In this case, the proceeding starts by closing all open power switches, one after the other, so that the best power flow path can be established by using many approximations. Borozan et al. [105] have introduced a similar method to [104]. However, this method has three main points:

- Estimation of the load;
- Effective determination of power losses in a network configuration;
- Cost-benefit analysis.

Civanlar et al. [106] have used a Branch Exchange technique heuristic method to determine the best configuration of a distribution system which reduced all active losses in lines. Baran and Wu [107] have tried to improve the Civanlar et al. [106] method by introducing two power flow formulas for approximation within the load handling systems.

Kashem et al. [108] have suggested the Branch Exchange technique method to the network reconfiguration. This is, basically, an extensive research method and it needs to take all the parameters of the switches into consideration. Cho and Chen [109] have analysed a schedule for the time reconfiguration. They studied hourly consumption during a year in order to define load conditions in each season. They used the Branch and Bound technique to obtain a network configuration which minimized active losses. Nara et al. [110] have suggested a distribution network reconfiguration method to reduce active losses using a genetic algorithm. Lin et al. [111] have improved the genetic algorithm applied to the network reconfiguration problem in order to reduce active losses in the lines.

Huang [112] suggested a genetic algorithm based on fuzzy set to the distribution network reconfiguration. Although some researchers [110-112] have proven the effectiveness of the genetic algorithm in the network reconfiguration, the solution's time is highly prohibitive. Lin and Chin [113-114] have introduced an algorithm for the distribution network reconfiguration based on the levels of tension, resistance and decision to determine the switches operation. Huang and Chin have suggested an algorithm based on the fuzzy operation to deal with the problem within the reconfiguration of the feeder. The goal in this approach is to minimize active losses and ensure that the load and the substations are balanced.

Liu et al. [116], Jung et al. [117] and Auguliaro et al. [118] have developed applications based on artificial intelligence to a configuration taking the minimization of losses into consideration. Hsiao [119] suggested a fuzzy multi-objective method to a network configuration based on evolutionary programming.

Matos et al. [120-122] have introduced a configuration method for radial distribution networks, contemplating a reduction of losses and service replacement using a meta-heuristic simulated annealing method. They suggest minimizing losses or unsupplied load. The total number of operations within the sectionalizing or the tie-switches is an objective function criterion. Other authors [121] have considered several loading scenarios, reaching one or several configuration solutions which have reduced energy losses as well as how many operations are executed by the tie-switches, with two goals within the work using the same meta-heuristic method.

Pereira et al. [123] introduced a model to identify the optimal operation strategy to minimize active power losses, considering transformer taps and capacitor banks operating in the distribution network. This problem was solved by the same meta-heuristic method and tested in a distribution network in Portugal.

Khodr et al. [124] introduced a mathematical model to reconfigure a 2344 nodes existing distribution network. This work used the mixed-integer linear programming (MILP) model. The goal is to search the distribution network's topology minimizing energy losses. Nowadays, the reconfiguration results obtained from this research are implemented in the above mentioned network which is now found running.

The problems of Distribution Systems Reconfiguration (DSR) and Optimal Power Flow (OPF) have been approached and studied separately by many authors in specialized literature, but these problems have not been studied within a single model using deterministic optimization techniques. Literature mentions reconfiguration without OPF [101–124], in which [125] the OPF is applied to the distribution system separately.

Gomes et al. [126] have introduced a heuristic approach to solve the DSR problem. Gomes [127] improves his heuristic strategy based on the OPF in which the state of each branch (opened/closed) appears through a continuous function. In this approach, all branches are initially considered as closed and from the OPF results, a heuristic technique is used in order to determine the next branch that is to be removed when opening a switch. Next, in the list of states of the branches, the opening is updated and the previous process is repeated until all branches are removed, thus making the system radial topology. This method does not ensure an optimal global solution to the RRDEE problem, as we will confirm in the next references.

Khodr et al. [128] have built a non-linear mathematical model to reconfigure the distribution network when there is a distributed generation production as well as considering load tap changer of the transformer and capacitor banks steps. In order to solve this mathematical model, the authors used the Benders decomposition technique in which the master problem is formulated as a mixed-integer quadratic programming (MIQP) model. The main function of this model is to obtain a radial network. The slave problem is formulated as a non-linear model in which the goal is for the network obtained through the master problem is send to the slave problem to execute an optimal power flow and therefore to adjust and check the technical constraints. Both problems (master and slave) are coupled by the Benders linear cut. The stopping criterion happens when the cost of the master problem's objective function equals the value of the slave problem's function. This methodology was tested in examples and real study cases taken from international literature. The execution time and number of iterations are accepted to recommend the use of this methodology as a real-time application. This algorithm was compared to several authors [104, 126, 127, 129, 130, 131].

The formulation presented can be applied in any distribution network, in any country and problem exposed is of extreme importance in order to reduce costs and it is still a challenge for future generations.

2.1. General Concepts

The reconfiguration technique may be considered of organizational type. When the studied circuit offers the possibility of interlinks with other nearby circuits, or if it is possible to feed some branches from different nodes within the same circuit, something which can be achieved through an operation work (open/close switches). This technique is only used when it is necessary to perform a new circuit configuration. To this end, it might be necessary to make some investment, for instance:

- Building new line sections which allow the connection;
- Installing drop-out fuses (shut-off equipment);
- Sectionalizing or switches/tie switches, etc.

The reconfiguration of electric power distribution networks is based on a load transfer from overload feeders of substations to lighter feeders' branch lines. The goal of this reconfiguration is to control the load level in branch lines involved in the operation, but also to improve voltage profiles throughout the branch lines, lessen total active losses and improve the quality of the electric service. The goal of the reconfiguration is to obtain an optimal topology as it is possible for a load of set of loads to be fed by a very long path or, in the worst conditions, with conductors of the smallest section, originating superior losses to those which would occur with other network configuration, leading to a power supplying of the loads through a flow line with better distribution conditions.

The problem of distribution networks reconfiguration can be solved through the Mixed-Integer Linear Programming (MILP) method. Most authors agree that the most representative mathematical model of this type of problem has at least two groups of variables: powers which circulate within the network (real and positive variables) and the decision to install lines or transformers within the network (binary variables) when necessary. Because of this, it is necessary to develop a mathematical model which can solve the problem of reconfiguration through a mixed-integer modelling.

The problem should be approached on an initial network which should be defined by the responsible engineer (planner), who usually has this type of information.

The distribution network is made of a set of L_i lines and another set of N_t, transformers/ substations which provide energy in an area obtained through N_d nodes of loads. These points are considered to have a specific geographic location, and known power consumption. Transformation points and connection lines may vary. Theoretically, there might be a line between two nodes and in each node, a transformer or a substation. In practice, there are technical rules, terrain blockage/obstacles and other causes which might limit how many connections are possible, transformation points and their ability which should be evaluated when there is network expansion.

Figure 2 (page 31) of the example case, shows the diagram of a 4 load nodes and 2-nodes transformation network with 12 lines which represent an initial network. In order to build this network, the following aspects were taken into consideration:

1. Transformers are located on the feed nodes which may represent a load centre node. In this node is considered without any connected loads. In this case of the passive distribution networks, the lines are only represented in one power flow direction only, which are installed from the transformers/power sources to the nodes of load they should feed. In the overhead distribution power-lines, transformers are located in a pole and they only exist as output lines, the connection between the transformer and the secondary circuit which it feeds.
2. Several transformers capacities are considered for the optimization problem obeying the rules and norms of the electric company or the equipment manufacturers. In all points, the desired capacities are evaluated. Transformers can be single-phased, bank of three single-phase or three-phased, according to the requirements of the circuit and the rules established by the companies involved. In Portugal, the transformers which were installed at transformation point are always three-phased.

3. Studying viable paths between the nodes, where there are no physical obstacles, such as buildings, gardens or other places where a line should not be built. These paths should have logical trajectories from a technical-economical point of view. For this type of nodes an extra line should be included if there are doubts as to the feeding point. All paths have two-way direction power flow lines, so that they can be fed by different transformers, substations or distributed power producers.
4. Initial localization of transformation points should be done according to certain logical criteria in the case of network expansion, imposed by the companies who own distribution networks.

2.2. Problem Statement

Electric power overhead distribution systems have mostly a radial configuration in order to facilitate inherent factors to protection, such as the coordination and attenuation of short-circuit currents to reduce equipment costs. Generally, it's possible to alter these systems' topology through by opening/closing switches, shut-off equipment and sectionalizing located in strategically places. Topology is altered, when keeping a radial topology, in such a way as to reduce active losses in the branches, improving the voltage profile for consumers, increasing the levels of reliability and eliminating and/or isolating faults as well as restoring the energy supply. Thus, these subsections are used for isolating defects and reconfigure the distribution network accordingly.

Distribution networks reconfiguration can be used as a planning and/or real time control tool of the system's operation. The online operation demands quick answers so that the right control actions can be taken within the systems' automatic command. In order to restore the service, an isolation of defective supply areas is carried out. When planning the operation, time spent in obtaining answers does not have such an important role as in the previous case. When planning the operation, there should be a search for configurations in order to obtain an optimal strategy of operation, minimizing active power losses, providing a good quality daily consumption (magnitude of the axial stress with a suitable profile, reliability, etc.), in addition to a cost reduction in operation and also within a network evolution perspective.

In summarise, it is possible to say that the reconfiguration problem within distribution networks consists of searching for an optimal operation strategy, in order to minimize active power losses in the branch lines and providing a suitable balance of loads within the three-phase feeders system, taking into account aspects such as reliability, protection and quality in energy supply to consumers/end users. In literature there are two approaches to this distribution network reconfiguration problem:

1. Using search techniques based on heuristics;
2. Using optimization techniques in which meshed networks are initially used, adopting the condition in which all switches are closed. Switches are successively opened in order to obtain a radial configuration. This optimization technique is usually used to solve the problem and is based on the Branch and Bound algorithm.

The exact solution to the problem of reconfiguration involves the selection, among all possible configurations, of that which has less active power losses. In this search phase, all possible configurations are tested in order to try and find the exact solution to the problem. However, it is difficult at a computational level, because the number of possible configurations obtained by the switches' position grows exponentially along with the number of switches position in the network, slowing the process in real systems. The exponential growth of the number of possibilities is called combinatorial explosion.

Because of this problem, approximate heuristic algorithms which might be suggested to solve the problem of reconfiguration do not assure an optimal global configuration, but only that the solution is good or almost optimal. Heuristic algorithms try to establish criteria which might be used to eliminate undesired options for the position of subsections in order to reduce the dimension of the problem.

Electric power distribution networks are mostly radially configured to reduce the costs arising from related protections and with the aim to facilitate operations within themselves, such as the coordination and attenuation of short-circuit currents and trying to reduce equipment costs. Generally, these networks allow a topology modification using switches located in strategic points. The aim of changing the topology while keeping the radial configuration is to reduce active power losses in branch lines, improve tension levels for consumers, increase reliability rates and eliminate or isolate faults, restoring power supply. Thus, switches or sectionalizing are used with two aims: isolating the defects and reconfigure or manage topology making it very flexible in operation.

A radial system has low short-circuit currents, simple protection equipment, low operational and building costs, but also low reliability rates. Because of this, distribution networks are built with switches in strategic points in order to help sectioning and reconfiguring the network. Generally, it is possible to say that the reconfiguration problem within distribution networks consists of searching for an optimal operation strategy, in order to minimize active power losses in the branch lines and providing a suitable balance of loads within the three-phase feeders system, taking aspects such as reliability, protection and quality in energy supply to consumers into account.

2.3. Secondary Aspects

The current view of energy systems includes some new technologies and rules, hard to picture some decades ago. Before these new realities which energy systems must face, changing the rules of the electric sector, in other words, of the production, transport and distribution sectors, and a growing competitiveness which requires more efficient means of exploration in economic terms.

In some countries, the energy production, transport and distribution sectors were taken by public companies and, in some cases, all these sectors were taken by a single company at a national level. In other countries, private companies or companies with partially private funds participated in this sector. However, in either situation, there was no competiveness among distribution companies and each consumer could be served by one single company, depending on his/hers localization. Thus, each distribution company owned the monopoly of its respective geographic area of operation. In this context, there was almost no competiveness within the electric sector.

Current demands for competiveness and the growing need for electricity have made it necessary for concessionary companies to obtain powerful, trustful and expandable tools in order to control their networks patrimony. The evolution of computer means and telecommunications has allowed more effective and thorough supervision, communication and control.

The evolution of distribution networks meant an evolution within the systems which motorized and controlled them. Nowadays, Supervisory Control and Data Acquisition (SCADA) control systems are used as a support to the motorization and control of these distribution networks. SCADA systems are distributed, counting on one or more Control Centres (CC) where specialized operators monitor and control the underlying process. SCADA systems are progressively becoming one of the most complex systems with the inclusion of new applications which aim to provide a better assistance to operators in order to take the right decisions with the support of optimization techniques and their real time applications. For instance, the use of Energy Management Systems (EMS) applications can be determinant for the operation of electric power distribution networks in the new context of the smartgrids.

The growth of these distribution networks, as well as a growing competiveness to which these economical activities have been subjected to in the last few years have created the need to obtain increasingly more effective help from operators at Control Centres, so that networks can be operated in schemes as optimized as possible and with minimal unavailability time. Thus, an optimal operation of distribution systems to reduce power and energy losses and therefore, to reduce generated power and emission of polluting substances has gained importance.

Worries over pollution have been gaining importance and allowing the creation of compromises which have progressively become rules and force a change on how pollution dispatch is carried out nowadays.

In technological terms, evolution includes, for example, the appearance of new control equipment, including Distribution Management System (DMS) applications based on open system architecture, projected according to internationally recognized standards and standardised. DMS products are independent of operating systems and work with most popular operating systems such as UNIX, Linux and Microsoft Windows™. The system's architecture is completely redundant in order to offer high availability in critical missions and provides real time high performance graphs. DMS products also work with Common Information Model (CIM), Electric Power Research Institute (EPRI), and Control Centre Application Program Interface (CCAPI).

The DMS product provides concessionary companies with a complete set of applications and tools for an efficient, trustworthy and economical management of distribution networks. Its sophisticated network model works with unbalanced three-phase networks and serves as a basis for all DMS applications. The DMS product provides interfaces with most Geographical Information System (GIS) systems for initial population and future online updates, both of the network model and of the operational screens. DMS applications are completely integrated to SCADA and EMS products and use a common real time and study mode database.

3. FORMULATION OF THE PROBLEM OF DISTRIBUTION NETWORKS RECONFIGURATION

The problem of distribution networks reconfiguration can be formulated as an MILP type optimization problem. Because of this, it is necessary to develop a type of methodology that can solve the problem under analysis. In order to formulate the problem, it is necessary to define a mathematical model so that annual costs of active power losses within the network can be minimized (small investments). Active power losses have a square expression, according to the power rate which circulates through the lines. Thus, it is important to conduct approximations and use linear optimization techniques which lead to exact mathematical solutions and reduced computer time calculation.

In order to approach distribution networks reconfiguration, a mathematical model was conceived to minimize annual costs of power losses on the lines in a case in which the company does not want to invest on building new line sections nor new substations. Model calculations can include has many aims as the operator/planner wishes.

Active power losses have a square expression according to the power which circulates on the lines. According to some authors [10], [42, 43], [46], [67] it is estimated that conducting approximations and using linear methods which offer almost exact mathematical solutions should be used.

The developed mathematical model has the following components:

Variables declaration: Includes two variables groups:

a) Representation of the powers which circulate through early network line segments. Powers occupy the first column (L_i) on the coefficient matrix of the mathematical model's variables and are real variables:

$$X_i \geq 0;\ i = 1, \dots, L_i$$

b) The variables for the decision of opening/closing the line selected, which are attached to correspondent powers, at a 1 to 1 ratio, in other words, powers which circulate through the lines if there are any. Decision variables occupy the following columns within the coefficient matrix:

$$X_i;\ i = L_i + 1, \dots, 2L_i$$

These are binary variables because they indicate when a certain line is opening/closing by means of a switch.

c) Decision variables which represent the need to expand or increase substations, for instance, installing *L* expansion capacities on a *p* point, are represented by the binary variable $Y_{P,L}$. If there are N_t feeding points to expand the network and each of them must evaluate several capacities of transformation points or substations to select the most suitable capacity (analysing technical-economic factors), the quantity will be different for each point. If it is $N_c(p)$, there are $\sum_{p=1}^{N_t} N_c(p)$ integer variables, occupying the columns:

$$i = 2L_i + 1, \ldots, 2L_i + 1 + \sum_{p=1}^{N_t} N_c(p)$$

3.1. Objective Function

The problem of electric power distribution networks reconfiguration can be formulated as a linear programming problem with binary variables. Its solution includes a selection of, among other possible configurations, the one which results in less active power losses and subjected to a set of constraints:

- Radial configuration of the distribution system;
- Acceptable voltage levels;
- Reliability, among others.

In this chapter, author is trying to minimize the costs of energy losses by using loss curves piecewise linearization to linear functions (straight small sections). Generally, this function can be written as follows:

$$Min\ F = Z_1 + Z_2 + Z_3$$

Objective function components are:

a) Cost of active power losses within the lines (Z_1)

In N_d load nodes, the set of lines which enter the nodes coincide with the total number of paths on the initial network. Considering the linearized expression of power losses to the maximum load condition [46] the following expression was obtained:

$$Z_1 = \sum_{i=1}^{N_d} \sum_{k=IN(i,1)}^{IN[i,Kl(i)]} C \cdot [S(k)]^2 \cdot W(i,k)$$

where:

$IN(i,k)$: Line reference number k which enters the i node; $k = 1, \dots, Kl(i)$

$Kl(i)$: Quantity of lines which enter the i node

$S(k)$: Is the apparent power which circulates through k lines (kVA); $k = 1, \dots, L_i$

C C : Is the coefficient of the cost of losses due to the power which circulates through the lines [*€/(kVA.year)*]

$C = C_1$ If there is a three-phase section branch and $C = C_3$ if is a single-phase of 3 conductors.

$$C_1 = \frac{\rho \cdot J_{eco} \cdot (12 \cdot Cd + Ce \cdot Fpd \cdot 8760) \cdot FVA \cdot D_{ij}}{\sqrt{3} \cdot U_l}$$

$$C_3 = \frac{4 \cdot \rho \cdot J_{eco} \cdot (12 \cdot Cd + Ce \cdot Fpd \cdot 8760) \cdot FVA \cdot D_{ij}}{\sqrt{3} \cdot U_l}$$

$$FVA = (1 + RI)^t; \quad t = 1, \dots, N$$

being:

FVA: Is the conversion factor for the current value of the losses cost, this is because the investments in several variants are made in different years. It is necessary to reduce costs in the first year [10]

ρ : Resistivity of the conductors or cable material $(\Omega - mm^2/km)$

Cd: Contracted power rate (*€/kVA/month*); is this type of electrical tariff fee is applied

Ce: Cost rate *(€/kVA.h*). It is supposed to be a uniform power factor to the network

Fpd: Loss factor

U_l: Line-to-line voltage (V)

R_l: Capital growth rate or simply, interest rate (%/ year)

N: Study years or lifetime of the project (year)

$D_{ij} = L$: Distance between two nodes (m)

The linearization of the objective function was conducted according to the Venikov criterion [46]. This approximation supposes that the facilities lines conductors are working close to the rated current which is the economic density (J_{eco}).

$$I = J_{eco} \cdot S_{ec}$$

where:

J_{eco}: Current economic density (*A/mm*^{2})

S_{ec}: Conductor's cross section (*mm*^{2})

The active power losses are:

$$\Delta P = k' \cdot R \cdot I^2 = k' \cdot R \cdot I \cdot I$$

replacing:

$$\Delta P = k' \cdot R \cdot J_{eco} \cdot S_{ec} \cdot I$$

and therefore:

$$I = k'' \cdot \frac{S}{U_l}; \qquad R = \frac{\rho \cdot L}{S_{ec}}$$

Where:

k' and k'' are constants which depend on the type of service (single-phase or three-phase).

S : Load power magnitude (kVA)

R : Conductor's resistance (Ω/km)

I : Current passing through the conductor (A)

Substituting:

$$\Delta P = \frac{k \cdot \rho \cdot L \cdot J_{eco}}{U_l} \cdot S$$

It is a linear expression is function of the apparent power.

b) Investment cost on the network lines (Z_2) in the case of expanding the network

The expression used to this end was:

$$Z_2 = \sum_{i=1}^{N_d} \sum_{k=IN(i,1)}^{IN[i,Kl(i)]} K_{el} \cdot W_{(i,k)}$$

Using the formula:

$$K_{el} = (E_n + N_{el}) \cdot K_{il}$$

where:

E_n: Economic effectiveness ratio (1/year).

$$N_{el} = (N_{al} + N_{oe})$$

N_{el}: Scanning the lines rule or utility norm (1/year)

N_{al}: Amortizing the lines rule or amortizing norm (1/year)

N_{oe}: Company's fixed costs rule (1/year)

K_{il}: Line investment costs (€/km)

$W_{(i,k)}$: Decision of installing a line from the *k* to the *i* node

c) Transformers investment cost (Z_3)

There are N_t feeding points with $N_c(p)$ evaluated capacities in each point.

$$P = 1, \dots, N_t$$

then:

$$Z_3 = \sum_{p=1}^{N_t} \sum_{l=1}^{N_c(p)} K_{et} \cdot Y_{(p,l)}$$

The K_{et} parameter includes a capital inflation rate, a fixed exploration costs rule and costs that come from iron losses in the transformers.

where

$$K_{et} = (E_n + N_{et}) \cdot K_{it} + K_{\Delta Fe}$$

where:

K_{it}: Total investment cost in the transformers (€)

$$N_{et} = (N_{at} + N_{oe})$$

N_{et}: Fixed exploration costs rule or exploration norm (1/year)

N_{at}: Transformers amortization rule or amortizing norm for transformer (1/year)

$K_{\Delta Fe}$: Cost that come from iron losses in the transformers (€/year)

$Y_{(p,l)}$: Binary variable for the decision of installing a *L* capacity on point *P*

The objective function includes copper loss costs in the transformers, taking another author [10] into consideration. Indicating the reasons for this cost:

- Load losses are expressed considering the square function.

$$\Delta P = P_k \cdot (S/S_n)^2$$

where:

Δp: Copper losses in the transformers (kW)

P_k: Power losses with loads (kW)

S : Apparent load power (kVA)
S_n: Apparent transformer nominal/rated power

In the objective function it is necessary to linearize the copper's loss expression, introducing small errors into the final results.

Transformers are limited to work between a maximum and minimum apparent power interval S (kVA), $S_{min}(l) \leq S \leq S_{max}(l)$ which is usually between the transformer's 70% and 100% rated or nominal capacity.
where:

L: Capacity of the installed transformer in each point (kVA)

Total loss costs have a square variation, and so the variable cost component can be used both in nominal capacity $S \leq S_n$ and smaller loads. This is a small component when compared to the fixed cost, concluding that:

$$(S/S_n)^2 < (S/S_n) < 1$$

and so, acceptable intervals to work on the variable cost difference, comparing several capacities, are small. When despised, there are no serious mistakes.

3.2. Technical Constraints

a) Power balance in the nodes (Kirchhoff's First Law for apparent power):

Includes all load nodes and is expressed according to the apparent power in (kVA), supposing that there is a single power factor to the network.

$$\sum_{k=1}^{K_1(i)} S[IN(i,k)] - \sum_{k=1}^{K_2(i)} S[IT(i,k)] = S(i); \quad i = 1, \ldots N_d; \quad k = 1, \ldots, K_2(i)$$

where:

$\mathrm{IT}(\mathrm{i}, \mathrm{k})$: Reference number for lines leaving the i node

S : Apparent power circulating through the line (kVA)

$K_2(i)$: Quantity of line leaving the *i* node
$K_1(i)$: Quantity of line entering the *i* node

$S(i)$: Apparent load power of the *i* node (kVA)

b) Capacity selection in each substation:

In each *p* transformation point, only one transformer already existent on the initial network will be installed (if necessary, in case of expansion) through the following expression:

$$\sum_{i=1}^{N_c(p)} Y(p,l) \leq 1; \qquad p = 1, \ldots, N_t; \qquad l = 1, \ldots, N_c(p)$$

c) Eliminating minimum load or underutilized substations:

This restriction helps eliminating disabled substations

$$\sum_{k=1}^{K_2(p)} S[IT(i,k)] \geq S_{min}(l) \cdot Y(p,l); \quad p = 1, \ldots, N_t;$$
$$l = 1, \ldots, N_c(p)$$

where:

$S_{min}(l)$: Minimum acceptable load limit imposed by the rules of related institutions (kVA) or electricity company utility.

d) Maximum load limitation or overload in substations:

This constraint enables a maximum load limitation in each substation, this avoiding their overload.

$$\sum_{k=1}^{K_2(p)} S[s, IT(i,k)] \leq S_{max}(l) \cdot Y(p,l); \quad p = 1, \ldots, N_t;$$

$$l = 1, \dots, N_c(p)$$

where:

$S_{max}(l)$: Maximum acceptable load limit (kVA). This value is usually defined by the electric utility.

e) Thermal limit in lines:

Power circulating in each line cannot exceed the maximum limit imposed by itself.

$$S(k) \leq SL(k)_{max} \cdot W(i,k); \qquad k = IN(i,k), \dots, IN[i, K_1(i)]; \\ l = 1, \dots, N_d$$

where:

SL_{max}: Maximum apparent power allowed by the line capacity (kVA)

$W(i,k)$: Binary variable of line decision

f) Radial configuration condition:

Each load node receives energy from one single line.

$$\sum_{k=IN(i,1)}^{IN[i,K_1(i)]} W(i,k) = 1; \quad i = 1, \dots, N_d$$

g) General load power limitation according to the capacities of substations or electric power generation unities. General power balance of the distribution system.

The total load power at a given transformation point must be lower or equal to the sum of the maximum capacity of each substation.

$$\sum_{k=IT(p,1)}^{IT[p,K_2(p)]} S(k) \leq S_{max}\,[N_c(p)] \cdot Y[p,N_c(p)];$$

$$p = 1, \dots, N_t$$

where:

$S(k)$: Apparent power sent in each output line capacity (kVA)

h) Eliminating two-way power flow directions on the lines:

The line should have one way only.

If the $\mathrm{lk} = \mathrm{IT}\,(\mathrm{i},\mathrm{k})$; $\mathrm{k} < \mathrm{K}_1(\mathrm{i})$ and $K_1 = IN(i,k)$; $k < K_2(i)$

then:

$$W(i,lk) + W(i,kl) \leq 1; \quad j = 1, \dots, N_d$$

where:

i : Node references

kl : Line references

The constraint on the maximum voltage drop is not directly established in this model. Initially, the network's configuration is not known, such as, for instance, the number of network branch lines. A branch line is a section that starts from the transformer and finishes on the last connected load. It should be pointed out that when network losses rise, voltage drops rise, and so the cost of power losses in the objective function demands a punishment in the cost minimization process [10-12], [16].

One way of confirming the demands of voltage drops is when the network is relatively large, is through the power flow which circulates through the radial system after finishing the optimization process. According to the results of the power flow, it is possible to confirm is the values have violated the capacity of line sections.

Constraints f and h might be considered redundant, but in practice it is possible to observe that they help increasing the convergence speed while searching for an optimal solution.

The quantity of variables considers slack and artificial variables through the following expression:

$$QV = 2 \cdot L_i + \sum_{p=1}^{N_t} N_c(p)$$

where:

QV : Quantity of mathematical model variables.

This quantity depends on the total number of lines (L_i), the quantity of transformation points (N_t) and suggested capacities in each point $[\mathrm{N_c(p)}]$.

Amount of expressed restrictions:

$$QR = L_i + 2 \cdot N_d + 2 \cdot N_t + NE + 2 \cdot \sum_{p=1}^{N_t} N_c(p)$$

where:

QR: Quantity of constraints on the mathematical model

NE : Number of equations resulting from two-way flow lines

3.3. Type of Problem and Identification of Optimization Techniques to its Solution

The problem of the distribution networks reconfiguration can be formulated as a MILP type problem to which the solution would be the optimal topology of the network. All technical constraints must be satisfied, such as the power balance in all the nodes and the feeders, minimum and maximum substations load limits, thermal limit in the lines and the radial configuration condition of resulting networks. This is a combinatorial optimization programming problem and has two types of variables: binary or complicated variable, as they are usually called which are decision and continuous variables and represent the powers circulating through the branches of the network, and decision binary variables of the substations which may be installed into the transformation points. When the number of these binary variables is increased, execution time becomes exponential function according to the time of computer calculation.

This type of problem is difficult to handle because of its combinatory nature, in addition to the difficulty in creating a mathematical formulation of certain constraints such as the radial configuration of resulting networks. There is also the problem with integer and continuous variables, which makes it a MILP problem even more difficult to solve.

The network on Fig. 1 has 4 load nodes and 2 nods as transformation /substation points. The network is made of two branch lines which are represented, in the figure, with thick lines. The dotted lines are open lines in normal operation conditions and their switches are represented on figure 1. The conductor parameters and technical data are shown on Table 1. Each circuit is fed through a 100 kVA transformer. Loads are shown in Fig. 1.

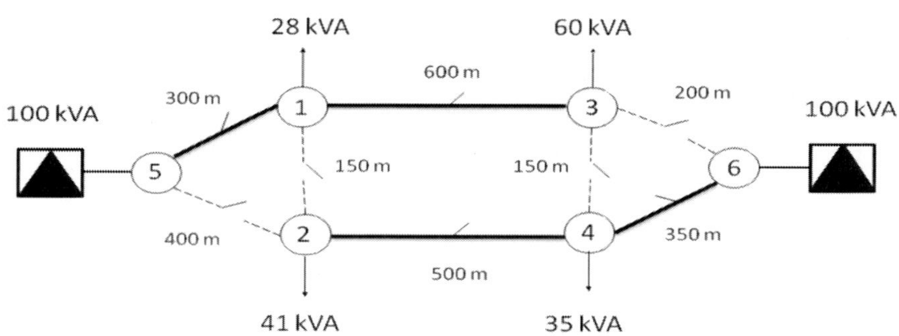

Figure 1. Current state distribution network.

It is necessary to know the current state of the network and, in order to obtain that information, a power flow for radial distribution networks is executed. In this case, it is very simple to know the power loss values. These values are shown on table 2 and, when they are analysed, it is possible to deduce that this network is not operating in an optimal conditions. In order to determine the state of the optimal operation of this network, it is necessary to formulate the optimization problem.

Table 1. Network data

Branch	Sending Node	Ending Node	Distance (m)	Conductor Section (mm²)	R (Ω/km)	X (Ω/km)	I_{MAX} (A)	J_{ECO} (A/mm²)
1	5	1	300	3x400 AL	0.102	0.096	515	0.5114
2	5	2	400	3x400 AL	0.102	0.096	515	0.5114
3	2	1	150	3x150 AL	0.265	0.101	285	0.8631
4	1	2	150	3x150 AL	0.265	0.101	285	0.8631
5	3	1	600	3x150 AL	0.265	0.101	285	0.8631
6	1	3	600	3x150 AL	0.265	0.101	285	0.8631
7	4	2	500	3x150 AL	0.265	0.101	285	0.8631
8	2	4	500	3x150 AL	0.265	0.101	285	0.8631
9	3	4	150	3x150 AL	0.265	0.101	285	0.8631
10	4	3	150	3x150 AL	0.265	0.101	285	0.8631
11	6	3	200	3x400 AL	0.102	0.096	515	0.5114
12	6	4	350	3x400 AL	0.102	0.096	515	0.5114

Table 2. Current state network losses

Current state of the network		
Branch	Power losses (kW)	Power losses (kVA)
5-1	1.48	1.69
1-3	3.58	3.89
2-4	1.39	2.22
4-6	1.29	1.70
Total	7.74	9.50

3.4. Analysing and Validating the Results

The optimization problem is based on the initial network shown on Fig. 2. In this figure, a single power flow sense is shown when there is no doubt about its sense, for instance, between nodes 5-1, 5-2, 6-3, and 6-4. Logically, power circulates from the source nodes until loading nodes where these powers will be consumed. When flow senses are not obvious, then they are represented with double opposite senses in order to give node 1 the possibility to, for instance, receive energy from the node 5 transformer or any other way through node 6 which represent another power source.

In order to formulate the optimization problem, powers are represented by the letter S and directions are represented by the numbers of involved nodes respectively. For instance, S_{51} means that the apparent power circulates from node 5 to node 1. This power is apparent and its unit is kVA.

In order to find the optimal operational configuration of the radial network, it is necessary to minimize power losses in this case, because circuits are built and so, the investment cost is null, and it is not necessary to build new branches.

The network in figure 2 has 12 lines and 2 transformers installed in nodes 5 and 6 respectively. Our aim is to minimize power losses which are the square function of the apparent power. It is necessary to calculate objective function coefficients. These values are shown in table 3.

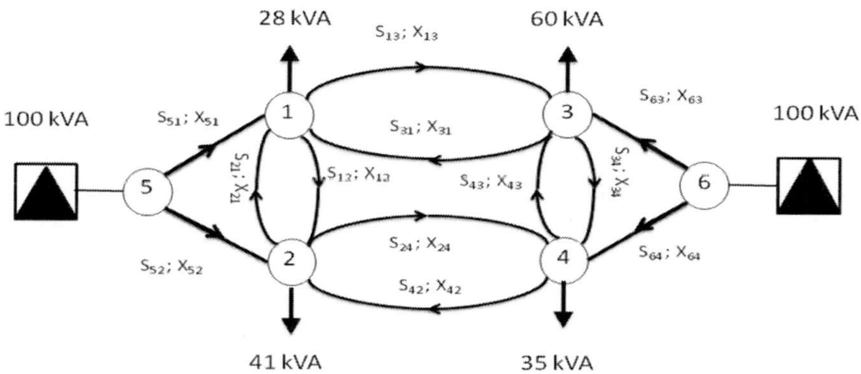

Figure 2. Initial considered network.

Table 3. Technical features of the conductors

Branch	Initial Node	Final Node	Section (mm^2)	Conductor Thermal capacity per phase (kVA)	Loss coefficient * 10^{-3} (kVA)
1	5	1	3x400 AL	118.45	19.21
2	5	2	3x400 AL	118.45	25.61
3	2	1	3x150 AL	65.55	16.21
4	1	2	3x150 AL	65.55	16.21
5	3	1	3x150 AL	65.55	64.85
6	1	3	3x150 AL	65.55	64.85
7	4	2	3x150 AL	65.55	54.04
8	2	4	3x150 AL	65.55	54.04
9	3	4	3x150 AL	65.55	16.21
10	4	3	3x150 AL	65.55	16.21
11	6	3	3x400 AL	118.45	12.81
12	6	4	3x400 AL	118.45	22.41

3.5. Mathematical Model of the Case under Study

3.5.1. Objective Function of the Case under Study

In order to formulate the objective function to minimize line power losses, it is necessary to know how many lines are there in Fig. 2. When observing this, it is possible to confirm that it has 12 lines and so, the objective function will have 12 terms as it is possible to confirm on the following equation:

$$MIN\ F = 10^{-3} * (19.21 * S_{51} + 25.61 * S_{52} + 16.21 * S_{21} + 16.21 * S_{12} + 64.85 * S_{31} + 64.85 * S_{13} + 54.04 * S_{42} + 54.04 * S_{24} + 16.21 * S_{34} + 16.21 * S_{43} + 12.81 * S_{63} + 22.41 * S_{64});$$

3.5.2. Technical Constraints of the Case Study

1. Power balance in the nodes (Kirchhoff's First Law):
   It includes all loading nodes and is expressed according to the apparent power in (kVA), assuming a single power factor to the network. In this network, there are 4 loading nodes, which lead to the following 4 equations:

   $$S_{31} + S_{21} + S_{51} - S_{13} - S_{12} = 28;$$

   $$S_{52} + S_{12} + S_{42} - S_{21} - S_{24} = 41;$$

   $$S_{63} + S_{13} + S_{43} - S_{31} - S_{34} = 60;$$

   $$S_{64} + S_{34} + S_{24} - S_{42} - S_{43} = 35;$$

2. Selection of the capacity of each substation.
   In this network there are 2 nodes with existing transformation points, so the binary variables are fixed with the optimization process.

   $$Y_1 = 1;$$

   $$Y_2 = 1;$$

3. Eliminating minimum load or underused substations.
   In the example there are 2 nodes with transformation points, so there will be 2 equations:

   $S_{51} + S_{52} - 50 * Y_1 >= 0;$

   $S_{63} + S_{64} - 50 * Y_2 >= 0;$

4. Maximum load limitation or overload in substations:
   In the example there are 2 transformation nodes, so there will be two equations:

   $S_{51} + S_{52} - 95 * Y_1 <= 0;$

   $S_{63} + S_{64} - 95 * Y_2 <= 0;$

5. Thermal limit in lines.
   The network has 12 lines, so there will be 12 equations.

   $S_{51} - 118.45 * X_{51} <= 0;$

   $S_{52} - 118.45 * X_{52} <= 0;$

   $S_{21} - 65.55 * X_{21} <= 0;$

   $S_{12} - 65.55 * X_{12} <= 0;$

   $S_{31} - 65.55 * X_{31} <= 0;$

   $S_{13} - 65.55 * X_{13} <= 0;$

   $S_{42} - 65.55 * X_{42} <= 0;$

   $S_{24} - 65.55 * X_{24} <= 0;$

   $S_{34} - 65.55 * X_{34} <= 0;$

   $S_{43} - 65.55 * X_{43} <= 0;$

$S_{63} - 118.45 * X_{63} <= 0;$

$S_{64} - 118.45 * X_{64} <= 0;$

6. Radial condition of operation of the network.
   In this example the network has 4 load nodes, so there will be 4 equations:

   $X_{51} + X_{21} + X_{31} = 1;$

   $X_{52} + X_{12} + X_{42} = 1;$

   $X_{13} + X_{63} + X_{43} = 1;$

   $X_{34} + X_{24} + X_{64} = 1;$

7. General power limitation according to the capacities of substations or electric power generation unites. General power balance of the distribution system. All the loads must be covered by the available generated power plus the power losses.
   Each example has 1 general network power balance equation:

   $S_{51} + S_{52} + S_{64} + S_{63} - 95 * Y_1 - 95 * Y_2 <= 0;$

8. Eliminating two-sense power flow on the lines. Between both paths there should be only one sense.
   In this example there are 4 two-sense arches, so there will be 4 equations:

   $X_{31} + X_{13} <= 1;$

   $X_{24} + X_{42} <= 1;$

   $X_{21} + X_{12} <= 1;$

   $X_{34} + X_{43} <= 1;$

   Binary variables are: X_{51}, X_{52}, X_{21}, X_{12}, X_{31}, X_{13}, X_{24}, X_{42}, X_{34}, X_{43}, X_{63}, X_{64}, Y_1, Y_2.

The necessary format to be executed in general optimization software LINGO is that shown in figure 3.

```
!;
! Objective function of the 4 load nodes, transformation nodes and twelve lines network;
!;
 MIN = (19.21 * S51 + 25.61 * S52 +
16.21 * S21 + 16.21 * S12 + 64.85 *
S31 + 64.85 * S13 + 54.04 * S42 + 54.04
* S24 + 16.21 * S34 + 16.21 * S43 +
12.81 * S63 + 22.41 * S64) * 10^-3 + 0 * y1 + 0 * y2;
!;
!;
! Restrictions;
!;
!;
! 1.Power balance in the nodes: Kirchhoff's First Law;
!;
 S31 + S21 + S51 - S13 - S12 = 28;
 S52 + S12 + S42 - S21 - S24 = 41;
 S63 + S13 + S43 - S31 - S34 = 60;
 S64 + S34 + S24 - S42 - S43 = 35;
!;
!;
! 2. Capacity selection in each substation;
!;
 y1 = 1;
 y2 = 1;
!;
!;
! 3. Eliminating minimum charge or underused substations;
!;
 S51 + S52 - 50 * y1 >= 0;
 S63 + S64 - 50 * y2 >= 0;
!;
!;
! 4. Maximum load limitation or overused substations;
!;
 S51 + S52 - 95 * y1 <= 0;
 S63 + S64 - 95 * y2 <= 0;
!;
!;

! Thermal limit in the lines;
!;
 S51 - 118.45 * X51 <= 0;
 S52 - 118.45 * X52 <= 0;
 S21 - 65.55 * X21 <= 0;
 S12 - 65.55 * X12 <= 0;
 S31 - 65.55 * X31 <= 0;
```

Figure 3. (continued).

```
S13 - 65.55 * X13 <= 0;
S42 - 65.55 * X42 <= 0;
S24 - 65.55 * X24 <= 0;S34 - 65.55 * X34 <= 0;
S43 - 65.55 * X43 <= 0;
S63 - 118.45 * X63 <= 0;
S64 - 118.45 * X64 <= 0;
!;
!;
! Radial conditions;
!;
X51 + X21 + X31 = 1;
X52 + X12 + X42 = 1;
X13 + X63 + x43 = 1;
X34 + X24 + X64 = 1;
!;
!;
! General load power limitation according to the capacities of substations or electric power generation unities. General power balance of the distribution system;
!;
S51 + S52 + S64 + S63 - 95 * y1 - 95 * y2 <= 0;
!;
!;
! Eliminating two-way power flow on the lines. Between both paths there should be only one way!;
X31 + X13 <= 1;
X24 + X42 <= 1;
x21 + X12 <= 1;
X34 + x43 <= 1;
!;
!;
! Declaração das variáveis Binárias ou de complicação;
!;
@BIN( X51);
@BIN( X52);
@BIN( X21);
@BIN( X12);
@BIN( X31);
@BIN( X13);
@BIN( X24);
@BIN( X42);
@BIN( X34);
@BIN( X43);
@BIN( X63);
@BIN( X64);
@BIN( y1);
@BIN( y2);
!;
```

Figure 3. Problem formulation code in LINGO software format.

Software results (LINGO) just as they can be seen in figure 4.

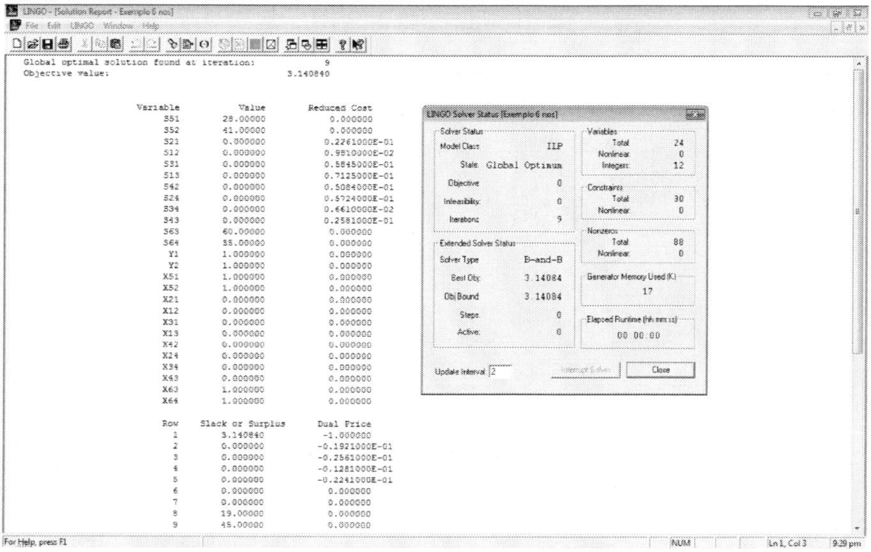

Figure 4. Execution of the mathematical model on LINGO Software.

LINGO software results can be seen in table 4.

Table 4. Results of the general optimization program

Global optimal solution found at iteration: 9		
	Objective value:	**3.140840**
Variable	**Value**	**Reduced Cost**
S51	28.0000	0.000000
S52	41.00000	0.000000
S21	0.000000	.2261000E-01
S12	0.000000	0.9810000E-02
S31	0.000000	0.5845000E-01
S13	0.000000	0.7125000E-01
S42	0.000000	0.5084000E-01
S24	0.000000	0.5724000E-01
S34	0.000000	0.6610000E-02
S43	0.000000	0.2581000E-01
S63	60.00000	0.000000
S64	35.00000	0.000000
Y1	1.000000	0.000000
Y2	1.000000	0.000000
X51	1.000000	0.000000
X52	1.000000	0.000000
X21	0.000000	0.000000

Table 4. (Continued)

Global optimal solution found at iteration: 9		
	Objective value:	**3.140840**
Variable	**Value**	**Reduced Cost**
X12	0.000000	0.000000
X31	0.000000	0.000000
X13	0.000000	0.000000
X42	0.000000	0.000000
X24	0.000000	0.000000
X34	0.000000	0.000000
X43	0.000000	0.000000
X63	1.000000	0.000000
X64	1.000000	0.000000
Row	**Slack or Surplus**	**Dual Price**
1	3.140840	-1.000000
2	0.000000	-0.1921000E-01
3	0.000000	-0.2561000E-01
4	0.000000	-0.1281000E-01
5	0.000000	-0.2241000E-01
6	0.000000	0.000000
7	19.00000	0.000000
8	45.00000	0.000000
9	26.00000	0.000000
10	0.000000	0.000000
12	90.45000	0.000000
13	77.45000	0.000000
14	0.000000	0.000000
15	0.000000	0.000000
16	0.000000	0.000000
17	0.000000	0.000000
18	0.000000	0.000000
19	0.000000	0.000000
20	0.000000	0.000000
21	0.000000	0.000000
22	58.45000	0.000000
23	83.45000	0.000000
24	0.000000	0.000000
25	0.000000	0.000000
26	0.000000	0.000000
27	0.000000	0.000000
28	26.00000	0.000000
29	1.000000	0.000000
30	1.000000	0.000000
31	1.000000	0.000000
32	1.000000	0.000000

3.5.1. Interpretation of the Results of the Case under Study

The values to be interpreted are in table 5 and the meaning of each of the variables is shown in figure 5. In this figure, there are 2 branch lines with different configurations from the current state because of the result of the optimization process. These branch lines have fewer power losses as shown in table 6.

In order to compare the optimization results, it is necessary to calculate branch lines power losses or circuits arising from the optimization, so that this might be accessed. As this is a simple network, then it is very simple to know how many power losses there are. These values are shown in table 6.

Table 5. LINGO software results

Global optimal solution found at iteration: 9		
	Objective value:	3.140840
Variable	Value	Reduced Cost
S51	28.00000	0.000000
S52	41.00000	0.000000
S63	60.00000	0.000000
S64	35.00000	0.000000
Y1	1.000000	0.000000
Y2	1.000000	0.000000
X51	1.000000	0.000000
X52	1.000000	0.000000
X63	1.000000	0.000000
X64	1.000000	0.000000

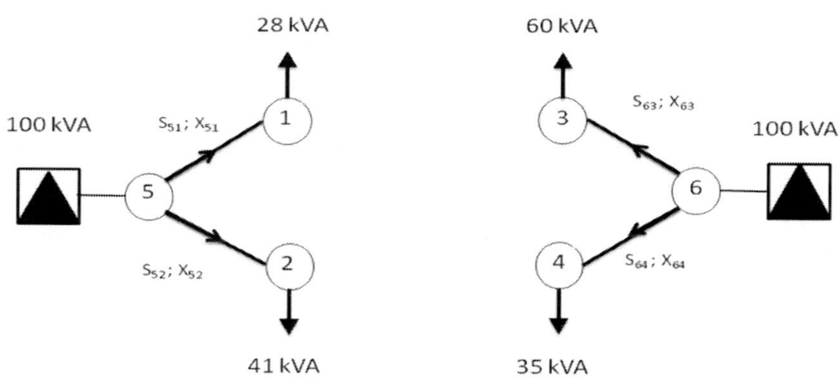

Figure 5. Network topology after reconfiguration.

Table 6. Network power losses after the reconfiguration

Branch		After reconfiguration	
Initial Node	Final Node	Losses (kW)	Losses (kVA)
5	1	0.15	0.54
5	2	0.43	1.05
6	3	0.46	0.77
6	4	0.27	0.78
Total		1.31	3.14

Comparing the results from the point of view of network power losses before reconfiguration and after this process, which are shown in table 7 (it contains losses values of apparent power in kVA and active power in kW), that is, of the current network state and the reconfiguration state. To sum up, network reconfiguration allows us to conclude that this method has many benefits, saving almost 66% of apparent power, allowing us to conclude that this method has great advantages when applied in distribution networks.

Table 7. Comparing losses in the current state and after reconfiguration

Branch	Current State		Reconfiguration	
	Losses (kW)	Losses (kVA)	Losses (kW)	Losses (kVA)
5-1	1.48	1.69	0.15	0.54
1-3	3.58	3.89	-	-
2-4	1.39	2.22	-	-
4-6	1.29	1.70	0.27	0.78
5-2	-	-	0.43	1.05
6-3	-	-	0.46	0.77
Total	7.74	9.50	1.31	3.14
Reduction	-	-	83.1 %	66.9%

3.6. Real Study Case

The methodology that has been explained in the previous section, was applied [20, 23, 25] to a real distribution network. This network has 3 substations. Their general data is shown in tables 8, 9 and 10.

Table 8. General Substation Data I (S/E)

Circuit	Length (m)	Transformers	Installed kVA
1	14181.1	243	11157.5
2	15401.7	376	17732.5
3	10026.7	258	13574
4	11505.0	307	16164
5	16008.2	179	10062.5
6	30021.7	352	15042.6
7	7361.0	12	197.5
8	2393.8	5	85

Substation 1 has an installed capacity in power transformers of 84.02 MVA, 8 circuits at 13.8 kV and has a total length of 106.9 km.

Table 9. General Substation Data II

Circuit	Length (m)	Transformers	Installed kVA
9	12829.2	169	9980
10	10376.1	105	8836.5
11	15245.33	226	8902.5
12	10386.9	230	9100
13	15895.34	104	9692.5

Substation II has an installed capacity in power transformers of 46.5 MVA, 5 circuits at 13.8 kV, and has a total length of 64.7 km.

Table 10. General Substation Data III

Circuit	Length (m)	Transformers	Installed kVA
14	20926	169	4902.5
15	3683	6	117.5
16	3712	13	600

Substation III has an installed capacity in power transformers of 5.6 MVA, 3 circuits at 13.8 kV, and has a total length of 28.3 km.

A database was created from these plants in which necessary values and parameters are as follows: 2799 nodes for the 14 circuits in 13.8 kV with the following data:

1. Nominal or rated voltage level in kV;
2. Average measured voltage level in kV;
3. Installed kvar in each circuit;

4. Type of dominating pole;
5. Load curve in each circuit in kVA and kW;
6. Sending node and ending node in each branch section after defining the nodes;
7. Length of each section in meters;
8. Conductor size in each section;
9. Phase number;
10. Total number of kVA installed in each node.

The process of defining nodes was conducted through the plans provided in a digital format and in which the following points of interest were considered:

1. Transformers banks;
2. Conductor's section or size;
3. Switches and isolators;
4. Transformers or capacitor banks;
5. Any other interest point in the circuit.

Measurements were also conducted in each circuit, more precisely on substations outputs, at the start of each circuit so that it would be possible to calculate the factors which define each of them.

Figure 6 represents an example of a load curve:

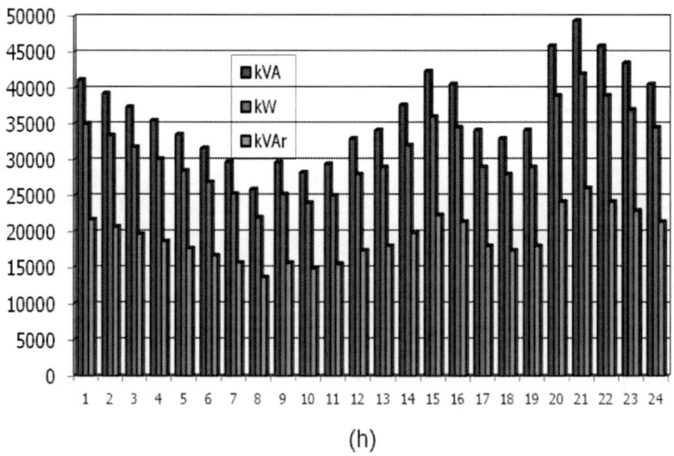

Figure 6. Typical load curve in a circuit.

3.6.1. Evaluation of the Current State in Each Circuit

In order to know the state of each circuit, it is necessary to carry out a power flow to radial distribution networks. The power factor was measured in each substation and the average value is 0.85.

After evaluating the current state in each circuit, power flow results and operation features in each circuit are shown in tables 11, 12, 13, 14, 15, 16, 17 and 18:

Table 11. Maximum load in S/E I circuits

Circuit	Installed kVA	Max kW	Capacity Factor
1	11157.5	9911	1.00
2	17732.5	15073	1.00
3	13574	11506	1.00
4	16164	13739	1.00
5	10062.5	8553.1	1.00
6	15042.6	12786	1.00
7	197.5	167.9	1.00
8	85	72.3	1.00

Table 12. Active power losses in S/E I circuits

Circuit	Line Losses		Transformer Losses		Total Losses	
	kW	In %	kW	In %	kW	In %
1	691.66	6.98	168.626	1.70	860.286	8.68
2	879.86	5.84	255.465	1.69	1135.325	7.53
3	616.17	5.36	175.857	1.53	792.027	6.89
4	618.70	4.50	231.628	1.69	850.328	6.19
5	540.98	6.32	147.796	1.73	688.776	8.05
6	724.83	5.67	221.189	1.73	946.019	7.40
7	0.21	0.12	3.311	1.97	3.521	2.09
8	0.01	0.01	1.405	1.94	1.415	1.95

Table 13. Voltage drop in S/E I circuits

Circuit	Length (m)	Worst Node	V (kV)	ΔV (%)
1	14181.1	187	11.65	11.71
2	15401.7	231	11.56	12.42
3	10026.7	212	11.72	11.22
4	11505 .0	155	12.10	8.33
5	16008.2	155	11.79	10.68
6	30021.7	198	11.57	12.37
7	7361.0	7	13.17	0.22
8	2393.8	9	13.20	0.03

Table 14. Energy losses in S/E I circuits

Circuit	Line Losses		Transformer Losses		Total Losses	
	MWh/year	In %	MWh/year	In %	MWh/year	In %
1	3535.84	5.52	862.03	1.35	4397.87	6.87
2	4497.67	4.61	1305.89	1.34	5803.55	5.95
3	3149.92	4.23	899.00	1.21	4048.92	5.44
4	3163.04	3.56	1184.17	1.33	4347.22	4.89
5	2765.54	5.00	755.55	1.37	3521.09	6.37
6	3705.48	4.48	1130.76	1.37	4836.24	5.85
7	1.07	0.10	16.92	1.58	17.99	1.68
8	0.05	0.01	7.17	1.41	7.22	1.42

Table 15. Maximum load in S/E II and III circuits

Circuit	Installed kVA	max kW	Capacity Factor
9	9980	7144.3	1.00
10	8836.5	7457.9	1.00
11	8902.5	6795.8	1.00
12	9100	7586.3	1.00
13	9692.5	8174.5	1.00
14	4902.5	4081.7	1.00
15	117.5	99.87	1.00
16	600	510	1.00

Table 16. Active losses in S/E II and III circuits

Circuit	Line Losses		Transformer Losses		Total Losses	
	kW	In %	kW	In %	kW	In %
9	282.32	3.95	119.218	1.67	401.538	5.62
10	308.13	4.13	98.639	1.32	406.769	5.45
11	138.39	2.04	134.234	1.98	272.624	4.02
12	156.22	2.06	130.573	1.72	286.793	3.78
13	743.35	9.09	136.392	1.67	879.742	10.76
14	54.95	1.35	73.275	1.80	128.225	3.15
15	0.04	0.04	2.257	2.26	2.297	2.30
16	0.31	0.06	8.842	1.73	9.152	1.79

Table 17. Voltage drops in S/E II and III circuits

Circuit	Length (m)	Worst Node	V (kV)	ΔV (%)
9	12829.2	136	12.33	6.56
10	10376.1	128	12.29	6.91
11	15245.33	171	12.67	4.01
12	10386.9	169	12.65	4.14
13	15895.34	84	11.21	15.06
14	20926	137	12.78	3.16
15	3683	4	13.19	0.06
16	3712	17	13.18	0.15

Table 18. Energy Losses in S/E II and III circuits

Circuit	Line Losses		Transformer Losses		Total Losses	
	MWh/year	%	MWh/year	%	MWh/year	%
9	23867.95	3.52	5818.98	0.86	29686.93	4.38
10	30080.86	3.67	8733.90	1.07	38814.75	4.74
11	21273.63	1.82	6071.57	0.52	27345.20	2.34
12	21362.55	1.84	7997.68	0.69	29360.23	2.53
13	18667.70	8.11	5100.02	2.22	23767.72	10.33
14	24812.61	1.2	7571.81	0.37	32384.42	1.57
15	7.25	0.04	114.27	0.63	121.51	0.67
16	0.34	0.05	48.44	7.03	48.78	7.08

3.6.2. Network Reconfiguration

The model which should be used in the network reconfiguration includes the following circuits:

1. Circuit 1;
2. Circuit 2;
3. Circuit 3;
4. Circuit 4;
5. Circuit 5;
6. Circuito 6;
7. Circuit 9;
8. Circuito 10;
9. Circuit 11;
10. Circuito 12.

Other circuits were not included in the reconfiguration study because they didn't have any connection or were close to other circuits/substations and had a low loss rate.

In order to carry out modification works in the network, a meeting with operators from the electric company was conducted in order to define possible connections or positions (opening/closing) of switches and a shorter connection distance of 55 meters was established because of the companies' economic restrictions. The file of possible connections is shown in table 19.

Table 19. Archive of possible connections agreed with the electric company

Sending Node	Ending Node	Length (m)	Conductor size	Initial Circuit	Final Circuit
35	143	1	4/0	1	2
56	181	44	2	1	2
66	249	1	1/0	1	2
77	146	43	4	1	2
136	151	1	2/0	1	9
139	195	24	2/0	1	2
213	124	1	1/0	2	10
217	98	34	2	2	12
231	104	39	4	2	3
244	153	1	4/0	2	10
29	8	35	1/0	3	4
42	53	43	2	3	4
55	98	41	4	3	4
40	83	1	4/0	1	3
90	132	37	4	3	4
70	97	1	2/0	1	3
113	74	45	2/0	3	12
118	10	1	2/0	3	11
119	68	1	2/0	3	12
133	121	1	2/0	3	10
140	110	1	2/0	3	10
146	105	5	2	3	10
191	97	1	4	3	10
210	131	1	2/0	3	10
24	128	1	2/0	4	5
40	123	23	4	4	5
186	95	1	4/0	4	12
189	84	1	1/0	4	12
202	134	1	2	4	11
199	130	43	2	4	11
13	37	13	1/0	5	6
176	196	1	2	5	11
10	263	1	1/0	5	6
25	38	1	4/0	9	10
27	22	10	1/0	9	10
40	26	1	1/0	9	10
60	72	26	1/0	9	10
80	83	10	1/0	9	10
106	87	41	2	9	10
112	146	26	4	11	11
135	138	52	2	11	12

After creating a database (2344 circuit nodes were included in the reconfiguration process), an optimization program was executed of which the results are that the switches and sectionalizing should be open. These results are shown in table 20.

Table 20. Circuits should be open

Node	Circuit	Node	Circuit
348	4	91	5
143	4	168	11
260	5	259	6
32	9	33	10
31	9	28	10
37	9	38	9
98	10	99	11
114	11	118	11
35	1	143	2
41	4	122	4
189	12	84	12
60	9	72	10

After the reconfiguration, the circuit database was changed mainly in order to study these circuits which are a result of the optimization process. In order to know the features and parameters of optimized circuits, a power flow was executed in radial networks to calculate losses in branches and voltage drops in busbars. The results of the configured network are shown in tables 21, 22 and 23.

Table 21. Active power losses in reconfigured circuits

Circuit	Total losses			
	Current State		Reconfigured State	
	kW	In %	kW	In %
1	860.286	8.68	503.22	6.48
2	1135.325	7.53	658.00	5.47
3	792.027	6.89	785.00	5.63
4	850.328	6.19	556.91	5.33
5	688.776	8.05	358.84	5.55
6	946.019	7.40	571.49	6.08
9	401.538	5.62	353.53	4.95
10	406.769	5.45	408.75	5.48
11	272.624	4.02	402.91	5.92
12	286.793	3.78	345.12	4.55

Table 22. Energy losses in reconfigured circuits

Circuit	Total losses			
	Current State		Reconfigured State	
	In MWh/year	In %	MWh/year	In %
1	4397.87	6.87	2572.48	5.13
2	5803.55	5.95	3363.70	4.33
3	4048.92	5.44	4012.76	4.45
4	4347.22	4.89	2846.85	4.21
5	3521.09	6.37	1834.38	4.38
6	4836.24	5.85	2921.46	4.81
9	2710.50	5.01	2750.24	4.63
10	2720.51	4.84	1926.24	4.52
11	1841.22	3.59	6383.39	5.12
12	1936.94	3.38	3781.12	4.54

Table 23. Voltage drops in reconfigured circuits

Circuit	Voltage Drop					
	Current State		Reconfigured State		Reduction	
	V (kV) in the worst node	%	U (kV) in the worst node	In %	V (kV)	In %
1	11.65	11.71	12.25	7.16	0.60	4.55
2	11.56	12.42	11.92	9.70	0.36	2.72
3	11.72	11.22	12.26	7.10	0.54	4.12
4	12.10	8.33	12.35	6.46	0.25	1.87
5	11.79	10.68	12.34	6.50	0.55	4.18
6	11.57	12.37	11.9	9.86	0.33	2.51
9	12.33	6.56	12.39	6.17	0.06	0.39
10	12.29	6.91	12.38	6.18	0.09	0.73
11	12.67	4.01	12.68	3.96	0.01	0.05
12	12.65	4.14	12.6	4.52	-0.05	-0.38

In order to evaluate operational features in the circuits after its reconfiguration and to quantify power and energy saving, as well as the improvement on voltages, it is necessary to compare the values before and after the reconfiguration. Results concerning power and energy saving are shown in tables 24 and 25.

Table 24. Power saving

Circuit	Reduction of power losses	
	Total Reduction	
	In kW	In %
1	-357.062	-2.20
2	-477.325	-2.06
3	-7.028	-1.26
4	-293.421	-0.86
5	-329.937	-2.50
6	-374.528	-1.32
9	-48.004	-0.67
10	+1.983	+0.03
11	+130.287	+1.90
12	+58.324	+0.77
Total	-1696.711	-8.17

Table 25. Energy saving

Circuit	Energy saving	
	Total Reduction	
	MWh/year	In %
1	-1825.387	-1.74
2	-2439.858	-1.62
3	-36.163	-0.99
4	-1500.364	-0.68
5	-1686.707	-1.99
6	-1914.776	-1.04
9	+39.740	-0.38
10	-794.268	-0.32
11	+4542.167	+1.53
12	+1844.173	+1.16
Total	-3771.444	-6.07

It is necessary to explain that during the reconfiguration process, some circuits with heavier load will lose some of it. This load will be transferred to other circuits with a lighter load. Thus, it is possible to conclude that the first case, circuits will lessen its losses and, consequently, voltage in busbars will improve. On the second case, circuits will increase its losses and its voltage will worsen. Generally, total losses will lessen and voltage will comply with the equipment's acceptable limits.

In tables 24 and 25, the signs (+) and (-) mean increase and decrease respectively. In these tables it is possible to confirm that, generally, reconfiguration leads to an approximately 6% to 8% saving in energy and power respectively without investment.

3.6.3. Perspectives for Electric Power Distribution Network Reconfiguration

The problem with electric power distribution networks is based on the determination of a radial topology and network interlink through the definition of the states (opened or closed) of switches engaged to distribution systems [102]. This topology optimizes a certain performance rate of the system, such as the reduction of technical power losses, taking operational constraints into consideration [132]. Subsection devices include: (*i*) usually closed switches (*NC*) and *(ii)* usually opened isolators (*NO*).

This option leads to a combined *Mixed-Integer Non-Linear Programming* (MINLP) problem because of the switching isolators. Considering the size of medium and great REPDN, searching for optimal solutions is difficult, because it requires evaluating a greater number of alternatives. The radial and interlinked feature of the distribution network is another aspect which makes this problem even more complex.

In literature there are methodologies which use discreet optimization techniques like *Branch and Bound* [103]. However, because of the combinatorial nature of the reconfiguration problem, this method requires analyzing a great number of configurations for real systems [31, 134]. Algorithms based on meta-heuristics like *Simulated Annealing* [134], *Ant-colony System* [135] and *Tabu Sereach* [136] have also been explored as a solution for this problem, as well as algorithms based on intelligent systems like *Genetic Algorithms* [110, 137], *Atificial Neronal Networks* [138] and methods using *Fuzzy Logic* [101, 139].

Several heuristic methods have been suggested as a solution technique to the REPDN reconfiguration *problem*, aiming at the minimization of network power losses [91,104, 130, 131, 140]. References [106, 107, 141] used an heuristic technique called *Branch-Exchange* to determine minimal active power losses configurations. These works introduce simplified strategies to calculate losses and reduce the search space, aiming at reducing computer efforts. In [142] a heuristic algorithm was suggested to minimize active power losses and load balance. The suggested method is based on the maximum voltage drop level in the branches. Gomes et al. [126] introduced an algorithm based on a heuristic strategy which sequentially opens the switches from a loop configuration, aiming at a greater loss growth in the REPDN. This method requires calculating non-linear power flow for each switch opening. In the continuation of this work, Gomes et al. [127] suggested using another strategy based on the Optimal Power Flow (OPF) to reduce the number of power flow simulations and, consequently, the computer effort.

Because of this problem's dimension and the computer time involved, few works about reconfiguration consider more than one system upload [109, 143, 144, 145]. Most works consider only one uploading period where the solution discovered is used in other hours throughout the day. So, it is possible to see that the system does not have an optimal network configuration in all uploading levels. Following this search line, this work presents an algorithm for the optimal distribution system reconfiguration taking different uploading levels into consideration. In the suggested methodology, switches are modelled according to continual functions and different load levels are incorporated into a single optimal power flow problem [146]. The algorithm considers an initial loop configuration and conducts the sequential opening of switches, until a radial configuration is obtained. Each opening is based on the sensibility of several switching discrete variables, obtained through *Lagrange* multipliers. To OPF solutions, in each algorithm step, the Primal-dual interior-point method is used [147, 148]. Results obtained are compared in order to evaluate the efficiency of the suggested method.

Khodr et al. [128] built a non-linear mathematical model to reconfigure the distribution network with distributed power production and considering the transformers' Load Tap Changer (LTC) and capacitor banks steps. In order to solve this mathematical model, authors used Benders decomposition techniques in which the master problem is formulated as a Mixed Integer Quadratic Programming model.

CONCLUSION

The main conclusions to be taken from this work are:

- Detailed explanation of the optimization mathematical formulation, applied to problems in the reconfiguration of electric power distribution networks;
- Application of the formulation in real cases of electric power distribution networks, controlling all variables through a MILP type optimization model. Its solution is based on the Branch and Bound

method. It may also be applied in other systems such as gas, water and communication distribution systems;

- Introduction of mathematical simplifications on a non-linear problem in order to linearize it, so that linear optimization techniques can be applied;
- Development of an optimization algorithm which has the advantage of being able to be used by engineers as a learning tool, exploring optimization problems in electric energy distribution networks;
- This model can be used by the distribution system operator, in the SCADA system to solve the system overload or to isolate the fault. Also it can be used by the SCADA and with the automated switches to take intelligent decision in the SCADA making the distribution networks more intelligent, acting as a smart network.
- The advantages of applying suggested methodology in the real case were an energy saving of about 3771.444 MWh/year, allowing an economic saving of 456721.87 €/year, considering a 0.1211 €/kWh cost without investments, only by opening/closing switches and shut-off devices.

Further Work to Enroll in Upcoming Activities

Following this work, a non-linear mathematical model to reconfigure distribution networks with distributed power production is to be building, considering LTC transformers and capacitor banks steps.

Benders *decomposition method* will be used in order to decompose the problem into 2 sub-problems, master and slave, to make its solution easier with acceptable computer times.

Additional Aspects

The practical aspect of this work should be complemented with a set of other lab activities, a study visit, which would be very useful for the trainee engineers or students to see the real world problems. This study visit should be to a distribution network switching centre (SCADA), so that students or beginner engineers can understand what actually happens, in other words, to understand operations conducted by the electric network operator as well as applying knowledge gathered during Lessons as a support tool to taking decisions in switching centres.

Occasionally, in the case of the students, a few lectures should be conducted in which specialists in some of the subjects approached in this curricular unit should participate.

Study visits and lectures should be conducted according to the availability of involved companies and students. This set of activities is not only related to the Lesson that has been presented nor with the curricular unit in which it is included and it should include a wider set of activities conducted in order to complement students' and beginner engineers training in the area of electric power distribution networks operation.

ACKNOWLEDGMENT

The author acknowledges the effort which has been done by Engineer Jonathan Exposito that has made reality this Book Chapter.

REFERENCES

[1] Gönen T., *"Electrical Power Distribution System Engineering"*, Textbook, CRC Press, Second Edition, Dec. 2007.

[2] Pansini J.A., *"Electrical Distribution Engineering*", Textbook, CRC Press, Third Edition, Dec. 2006.

[3] Kersting W.H., "*Distribution System Modeling and Analysis*", Textbook, CRC Press, 2006.

[4] Grigsby L.L., *"Electrical Power Generation, Transmission and Distribution"*, Textbook, CRC Press, Second Edition, 2006.

[5] Khodr H.M., Ostos D., Mora B., Urdaneta A.J., Yusta J.M., Paiva P., Hussein H., "Optimal reconfiguration in large electrical power distribution networks", *Proceedings of the IV Venezuelan Congress of electrical engineering, Caracas, Venezuela,* Set. 2004.

[6] Khodr H.M., Urdaneta A.J., Gómez J.F. y Ocque L.A., *"Modelo para la Optimización de Redes Secundarias de Distribución Utilizando algoritmos Híbridos"*. VIII Encuentro Regional Latinoamericano de la CIGRE Ciudad del Este, Paraguay, 03 de Jun. de 1999.

[7] Khodr H.M., Urdaneta A.J., Gómez J.F. y Ocque L.A. "Modelo para la Optimización de Redes Primarias de Distribución Utilizando Programación Lineal", *I conferencia Internacional de la región Andina y IEEE*, Margarita – Venezuela, 1999.

[8] Khodr H.M. *"Diseño óptimo de redes eléctricas de distribución secundaria"*, Tesis Doctoral, ISPJAE, Cuba, 1996.

[9] Khodr H.M., Gómez J.F., Barnique L., Vivas J.H., Paiva P., Yusta J.M. and Urdaneta A.J., "A Linear Programming Methodology For The Optimization Of Electric Power Generation Schemes", *IEEE Transactions on Power Systems,* vol. 17, Issue 3, Aug. 2002 pp. 864 – 869.

Additional References

[1] Mendoza F., Bernal-Agustin J.L., Dominguez-Navarro J.A., "NSGA and SPEA Applied to Multiobjective Design of Power Distribution Systems", *IEEE Transactions on Power Systems*, vol. 21, Issue 4, Nov. 2006, pp.1938-1945.

[2] Strausser E. A., "Computers Help Plan System*", Electrical World*, May 30, 1960.

[3] Bergougneux J., Lemoine J. C., Jomier R., *"Contribution à l'étude d'une politique optimale de developpement des réseaux de distribution",* E.D.F; R.G.E, Tome 84, no. 11, Nov. 1975, pp. 840-849.

[4] Chan K., "Computer-aided study and design in power system engineering using personal-computer-based application software", *Power Engineering Journal,* vol.5, no. 3, May. 1991, pp. 119-128.

[5] Hasselfield C. W., Wilson P., Penner L., "An automated method for least cost distribution planning", *IEEE Transactions on Power Delivery*, vol.5, no.2, Apr. 1990, pp. 1188-1194.

[6] Glamocanin V., "Optimal loss reduction of distribution networks", *IEEE Transactions on Power Systems,* vol. 5, pp.774-782, Aug. 1990.

[7] Barras J., Alec S. Pasche C., Chamorel P. A., Germond A. J., "Networks Simplex Method Applied to AC load-flow Calculation*", IEEE Transactions on Power Systems,* vol. PWRS-2, no. 1, Feb. 1987, pp. 197-203.

[8] Cespedes G. R., "New method for the analysis of distribution networks", *IEEE Transactions on Power Delivery*, vol. 5, no. 1, January 1990, pp. 391-396.

[9] Boardman J.T., Meckiff C.C., "A Branch and Bound Formulation of an Electricity Distribution Planning Problem", *IEEE Transactions on Power Apparatus and Systems,* vol. PAS-104, no. 8, Aug. 1985, pp. 2112- 2118.

[10] Fattal E., Armas A., Capri L., *"Modelo para la optimización de redes secundarias de distribución utilizando programación Entera-Mixta",* V Congreso de distribución de energía eléctrica, Venezuela, 1994.

[11] Gönen T., Foote B.L., "Distribution system planning using Mixed-Integer Programming", *IEE Proceedings, Part C,* vol. 128, no. 2, March 1981, pp. 70-79.

[12] Gönen T., Ramirez-Rosado I.J., "Optimal multi-stage planning of power distribution system", *IEEE Transactions on Power Delivery,* vol. PWRD-2, no. 2, Apr. 1987, pp. 512-519.

[13] Gönen, T., Ramírez-Rosado I.J., "Review of Distribution System Planning Models: A Model for Optimal Multistage Planning", *IEE Proceedings Part.* C, vol. 133, no. 7, 1986, pp. 397-408.

[14] Grawford D.M., Holt S.B., "A mathematical optimization technique for locating and sizing distribution substation, and deriving their optimal areas", *IEEE Transactions on Power Apparatus and Systems,* vol. PAS-94, no. 2, Mar./Apr. 1975, pp. 230-235.

[15] Group on Distribution System Design, "Bibliography on Distribution Automation 1969-1982", *IEEE Transactions on Power Apparatus and Systems,* vol. PAS-103, no. 6, June 1984.

[16] Hindi K.S., Brameller A., "Design of Low Voltage Distribution Networks: A Mathematical Programming Method", *Proceedings IEE,* vol. 124, no.1, pp. 54-58, Jan. 1977.

[17] Huang J.A., Galiana F.D., "An integrated personal Computer graphics environment for power system education, analysis and design", *IEEE Transactions on Power Systems,* vol. 6, no. 3, Aug. 1991, pp. 1279-1285.

[18] Kaplan, M., Braunstein A., "Contribution to the Determination of the Optimum Site for Substation", *IEEE Transactions on Power Apparatus and Systems,* PAS-100 (5), 1981, pp. 2263-2270.

[19] Knight, U.G., "The Logical Design of Electrical Networks Using Linear Programming Methods", *Proceedings IEE,* vol. 33A, 1960, pp. 306-319.

[20] Khodr H.M., *"Técnicas modernas de ingeniería aplicadas a las redes de distribución en la OBE Boyeros",* Informe Técnico, CIPEL, Abril, 1994.

[21] Khodr H.M. *"Diseño óptimo de redes eléctricas de distribución secundaria",* Tesis Doctoral, ISPJAE, Cuba, 1996.

[22] Khodr H.M. “Algoritmo para la selección económica de los transformadores de distribución”, *Revista Ingeniería Energética* no. 2, ISPJAE, La Habana, Mar. 1995.

[23] Khodr H.M., Carvajal R., “Modelo para la optimización de redes eléctricas de distribución primaria utilizando la programación lineal”, Revista Tecnológica, *Serie Electroenergética, MINBAS*, la Habana, Oct. 1996.

[24] Khodr H.M., Ayala D.V., “Modelo para la optimización de redes eléctricas de distribución secundaria utilizando la programación lineal”, *Revista Investigación Operacional,* Universidad de la Habana, Feb. 1996.

[25] Khodr H.M., Carvajal R. “Estudio de la red de distribución primaria a 4,16 kV de la zona de Boyeros”, *Revista Ingeniería Energética*, no. 1, ISPJAE, La Habana, Feb. 1997.

[26] Khodr H.M.,“Sistema para la planificación de redes de distribución primaria”, *Revista Ingeniería Energética*, no. 1, ISPJAE, La Habana, Feb. 1997.

[27] Khodr H.M., Urdaneta A.J., Gómez J.F., Ocque L.A., *“Modelo para la Optimización de Redes Secundarias de Distribución Utilizando algoritmos Híbridos”,* VIII Encuentro Regional Latinoamericano de la CIGRE Ciudad del Este, Paraguay, 03 de Jun. de 1999.

[28] Khodr H.M., Urdaneta A.J., Gómez J.F., Ocque L.A. *“Modelo para la Optimización de Redes Primarias de Distribución Utilizando Programación Lineal”*, I conferencia Internacional de la región Andina y IEEE, Margarita – Venezuela, 1999.

[29] Khodr H.M., Gómez J.F., Barnique L., Vivas J.H., Paiva P., Yusta J.M., Urdaneta A.J., “A Linear Programming Methodology For The Optimization Of Electric Power Generation Schemes”, *IEEE Transactions on Power Systems,* vol. 17, Issue 3, Aug. 2002, pp. 864 - 869

[30] Leung, L.C., Khator, S.K., Schnepp J.C., “Planning Substation Capacity under the Single-Contingency Scenario”, *IEEE Transactions on Power Systems,* PWRS-10 (3), 1995, pp. 1442-1447.

[31] Liu C., Lee S.J., Vu K., “Loss minimization of distribution feeders: Optimality and algorithms”, *IEEE Transactions on Power Delivery,* vol.4, no.2, Apr. 1989, pp. 1281-1289.

[32] Nara K., Satoh T., Kuwabara U., Aoki K., Kitagawa M., “Multi-Year Expansion Planning for Distribution Systems”, *IEEE Transactions on Power Systems*, PWRS, 6 (3), 1991, pp. 952-958.

[33] Nara K. et al., "Distribution system expansion planning by multi-stage branch exchange", *IEEE Transactions on Power Systems*, vol. 7, no. 1, Feb. 1992, pp. 208-214.

[34] Nara K., Kuwabara H., Kitagawa M., Ohtaka K., "Algorithm for Expansion Planning in Distribution Systems Taking Faults into Consideration", *IEEE Transactions on Power Systems*, PWRS-9 (1), 1994, pp. 324-330.

[35] Ponnambalam K., Quintana V.H. and Vannelli A., "A Fast Algorithm for Power System Optimization Problems Using an Interior Point Method", *IEEE Transactions on Power Systems*, PWRS-7, no. 2, May 1992, pp. 892-899.

[36] Ramírez-Rosado I.J., Adams R.N., "Optimal design of electric energy distribution systems: Multiobjective mathematical modeling", *Modeling Identification and control,* 11th International conference, Innsbruck, Feb. 1992.

[37] Ramírez-Rosado I.J., Gönen T., "Pseudodynamic planning for expansion of power distribution systems", *IEEE Transactions on power Systems*, vol. 6, no. 1, Feb. 1991, pp. 245-254.

[38] Ramírez-Rosado I.J., *"Planificación de Sistemas de Distribución de Energia Eléctrica Mediante Técnicas de Optimización Avanzadas"*, Secretariado de Publicaciones de la Universidad de Zaragoza, 1992.

[39] Ramírez-Rosado I.J., Alvarez-Bel C., "Applications of Multiobjective Models to Power-Distribution Systems Design", *International Journal of Power and Energy Systems*, 15 (2), 1995, pp. 54-60.

[40] Ramírez-Rosado I.J., Adams R.N., Gönen T., "Computer Aided Design of Power Distribution Systems: Multiobjective Mathematical Simulations", *International Journal of Power and Energy Systems*, 14, (1), 1994, pp. 9-12.

[41] Willis H.L., Northcote-Green J.E.D., Tram H.N., "Computerized Distribution Planning - DATA Needs and results with incomplete data", *IEEE Transactions on Power Delivery*, vol. PWRD, no. 4, Oct. 1987, pp. 1228-1235.

[42] Willis H.L., Northcote-Green J.E.D. "Comparison of several distribution planning techniques", *IEEE Transactions on PAS*, Jan. 1985, p. 1278.

[43] Willis H.L., Northcote-Green, "Comparison of Several Computerized Distribution Planning Methods", *IEEE Transactions on Power Apparatus and Systems*, PAS- 104 (1), 1985, pp. 233-240.

[44] Yuan Y.H., Chen J.L., "Distribution planning using a Knowledge-Based Expert System", *IEEE Transactions on Power Delivery*, vol. 5, no. 3, July 1990.

[45] Thompson, G.L., Wall D.L., "A Branch and Bound Model for Choosing Optimal Substation Locations", *IEEE Transactions on Power Apparatus and Systems*, PAS-100 (5), 1981, pp. 2683-2688.

[46] Venikov V.A., "Kibernetika Elektriczeskich Sistiem", Moskowa, 1974.

[47] Gangel M.W., Siken J.P., Smith J.A., Taylor H.V., "Optimized Distribution and Subtransmission Planning by Digital Computer. III - Primary Main Circuit Design", *IEEE Transactions on Power Apparatus and Systems*, vol. PAS-84, no. 12, December 1965.

[48] Hsu Y.Y., Hwu J., Liu S.S., Chen Y.W., Feng H., Lee Y.M., "Transformer and feeder load balancing using heuristic search approach", *IEEE Transactions on Power Systems*, vol. 8, no. 1, 1993.

[49] Ponnavaikko M., et al, "Distribution system planning through a quadratic Mixed-Integer programming approach", *IEEE Transactions on Power Delivery*, vol. PWRD-2, no.4, Feb.1986, pp. 1157-1163.

[50] Parton K.C. *"Experiences of loss minimizing in large distribution networks"*, Sponsored by the Eureka information service group Inc. And Electric light & Power, Bonaventure Conference Center, Fort Lauderdale, Florida, January 13-15, 1992, pp. 1-13.

[51] Taylor T., Lubkeman D., "Implementation of Heuristic Search Strategies for Distribution Feeder Reconfiguration", *IEEE Transactions on Power Delivery,* vol. 5, no. 1, Jan.1990, pp. 239-246.

[52] El-Kady M.A., "Computor-Aided Planning of Distribution Substation and Primary Feeders", *IEEE Transactions on Power Apparatus and Systems,* PAS-103 (6), 1984, pp. 1183-1189.

[53] Gönen T., *"Electric Power Distribution System Engineering*", Text, Editorial McGraw Hill Book Company, University of Missouri at Columbia, USA, 1986, p. 737.

[54] *IEEE Tutorial on Power Distribution Planning Sponsored by the Power Engineering Education, Committee Distribution Planning Working Groups.* E.H.O-361-6-PWR, 1992.

[55] McCall L.V., Chambers B.J., "Defining a distribution system for computer", *IEEE Transactions on Power Apparatus and Systems*, vol. PAS-102, no. 8, Aug. 1983, pp. 2665-2669.

[56] Papadopoulos M., Hatziargyriou N.D., "Graphics Aided Interactive Analysis of radial distribution networks", *IEEE Transactions on Power Delivery*, vol. PWRD-2, no. 4, Feb. 1986, pp. 1297-1302.

[57] Sun D.I., Farris D.R., Cote P.J., Shoults R.R., Chen M.S., "Optimal distribution substation and primary feeder planning via the fixed charge network formulation", *IEEE Transactions on Power Apparatus and Systems*, vol. PAS-101, no.3, Mar. 1982, pp. 602-609.

[58] Willis, H.L., Tram H.N., Powell R.W, "Substation Site and Capacity Selection Based on Diversity Maximization", *IEEE Transactions on Power Systems*, PWRS-2 (3), 1987, pp. 210-217.

[59] Luenberger D.G., *"Introduction to linear and nonlinear programming"*, Addison-Wesley, 1973.

[60] Stankovic A.M., Calovich M.S., "Graph oriented algorithm for the steady-state security enhancement in distribution networks", *IEEE Transactions on Power Delivery*, vol. 4, no. 1, Jan. 1989.

[61] Fawzi, T.H., Ali K.F., El-Sobki S.M., "Routing Optimization of Primary Rural Distribution Feeders*", IEEE Transactions on Power Apparatus and Systems,* PAS-101 (5), 1982, pp. 1129-1133.

[62] Latorre G., *"Modelos estáticos para la planificación a largo plazo de la red de transporte de energía eléctrica",* Universidad Pontificia Comillas, 1993.

[63] Marshall A.C., Boffey T.B., Green J.R., Hague H., "Optimal Design of Electricity Distribution Networks", *Proceedings IEE*, vol. 138, Pt. C, no. 1,1991, pp. 69-77.

[64] Mikic O.M., "Mathematical Dynamic Model for Long-Term Distribution System Planning", *IEEE Transactions on Power Systems*, PWRS-1 (1), 1986, pp. 34-40.

[65] Ponnavaikko, M., Rao K.S., "Optimal Distribution System Planning", *IEEE Transactions on Power Apparatus and Systems,* PAS-100 (6), 1981, pp. 2969-2977.

[66] Rao Nagendra P.S., "An extremely simple method of determining optimal conductor section for radial distribution feeders", *IEEE Transactions on Power Apparatus and Systems*, vol. PAS-104, no. 6, Jun. 1985, p. 1439.

[67] Tram H.N., Wall D.L., "Optimal Conductor Selection in Planning Radial Distribution Systems", *IEEE Transactions on Power Systems*, PWRS-3 (1), 1988, pp. 200-206.

[68] Youssef H.K., et al., "Novel Optimization Model for Long Range Distribution Planning", *IEEE Transactions on Power Apparatus and Systems*, PAS-1 04 (11), 1985, pp. 3195-3202.

[69] Partanen J., "A Modified Dynamic Programming Algorithm for Sizing Locating and Timing of Feeder Reinforcements", *IEEE Transactions on Power Delivery*, PWRD-5 (1), 1990, pp. 277-283.

[70] Youssef H.K., Hackam R., "Dynamic Solution of distribution planning in intermediate time Range", *IEEE Transactions on Power Delivery*, vol 3, no. 1, Jan. 1988, pp. 341-348.

[71] Holland J.H., *"Algoritmos genéticos: Programas que evolucionan, simulando en cierto grado la selección natural, alcanzan a resolver problemas complejos, que ni siquiera quienes lo crearon comprenden plenamente"*, Investigación y ciencia, Sep., 1992.

[72] Holland J.Y., *"Adaptation in natural and artificial systems"*, University of Michigan Press, Ann Arbor, Michigan, 1975.

[73] Bernal-Agustín J.L., *"Aplicación de Algoritmos Genéticos al Diseño Óptimo de Sistemas de Distribución de Energía Eléctrica"*, Tesis Doctoral, Universidad de Zaragoza, España, Ene. 1998.

[74] Grefenstette J.J., "Optimization of control parameters for genetic algorithms", *IEEE Transactions on Systems Man and Cybernetic*", vol. SMC-16, no. 1, Feb., 1986.

[75] Marín F.J., García F., Sandoval F., "Algoritmos genéticos: una estrategia para la búsqueda y la optimización ", *Informática y automática*, vol. 25, nos. 3 y 4, pp. 5-15, Nov. 1992.

[76] Miranda V., Ranito J.V., Proença L.M., "Genetic algorithms in optimal multistage distribution network planning", *1994 IEEE Summer power meeting, 94 WM 229-5,* Jan. 1994, pp. 1927-1933.

[77] Ramírez-Rosado I.J., Bernal-Agustín J.L., "Optimization of Power Distribution Network Design by Application of Genetic algorithm", *International Journal of Power and Energy Systems*, vol. 15, no. 3, 1995, pp. 104-110.

[78] Ramírez-Rosado I.J., Bernal-Agustín J.L., "Genetic algorithm applied to design of large power distribution system", *IEEE Transactions on Power Systems,* vol. 13, no. 2, May 1998, pp. 696-703.

[79] Ribeiro Filho J.L., Treleaven P.H., Alippi C., *"Genetic Algorithm programming environments"*, Computer, pp. 28-43, Jun. 1994.

[80] Rudolph G., "Convergence analysis of canonical genetic algorithms", *IEEE Transactions on Neural Networks*, vol. 5, no. 1, Jan. 1994.

[81] Khator S.K. and Leung L.C., "Power Distribution Planning: A Review of Models and Issues", *IEEE Transactions on Power Systems*, PWRS-12 (3), 1997, pp. 1151-1159.

[82] CIGRÉ TF., "Artificial neural networks for distribution systems: A literature Survey", *Engineering Intelligent Systems*, vol. 1, No. 3, pp. 133-158, Dec. 1993.

[83] Khodr H.M., Ibrahim H., Yusta J.M., Bernal-Agustín J.L., Domínguez-Navarro J.A., *"Previsión de la Demanda Eléctrica a Corto Plazo mediante Redes Neuronales"*, 7as Jornadas Hispano-Lusas de Ingeniería Eléctrica, Universidad Carlos III de Madrid, España, 2001.

[84] Grefenstette J.J., "Optimization of control parameters for genetic algorithms", *IEEE Transactions on Systems Man and Cybernetic*", vol. SMC-16, no. 1, Feb.1986.

[85] Hajek B., "A Tutorial survey for theory and applications of Simulated Annealing", *Proceedings of 24th Conference on decision and control*", Ft. Lauderdale, Fl., Dec. 1985, pp. 755-760.

[86] Johnson D., Aragón C.R., Megeoch L.A. Schevonc, "Optimization by Simulated Annealing: An experimental Evaluation Part I, Graph Partitioning", *Operation Research*, vol. 37, no. 6, Nov.-Dec. 1989, pp. 865-892.

[87] Johnson D., Aragón C.R., Megeoch L.A., Schevonc, "Optimization by Simulated Annealing: An experimental Evaluation Part II, Graph Coloring and Number Partitioning", *Operation Research*, vol. 39, no. 3, May-Jun. 1991, pp. 378-406.

[88] Goswami, S.K., "Distribution System Planning using Branch Exchange Technique", *IEEE Transactions on Power Systems*, PWRS-12 (2),1997, pp. 718-723.

[89] Kuwabara, H., Nara K., "Multi-Year and Multi-State Distribution System Expansion Planning by Multi-Stage Branch Exchange", *IEEE Transactions on Power Delivery*, PWRD-12 (1), 1997, pp. 457-463.

[90] Nara K. et al., "Distribution system expansion planning by multi-stage branch exchange", *IEEE Transactions on Power Systems*, vol. 7, no. 1, Feb. 1992, pp. 208-214.

[91] Peponis G.J., Papadopoulos M.P., "New Dynamic Branch Exchange for optimal Distribution Planning", *IEE Proceedings Generation and Distribution*, vol. 144, no. 3, May 1997, pp. 333-339.

[92] Dorigo M., Maniezzo V., Colorni A., "The ant system: Optimization by a colony of cooperating agents", *IEEE Transactions on Systems, Man and Cybernetic B*, vol. 26, no. 2, 1996, pp. 29-41.

[93] Dorigo M., Gambardella L.M., "Ant colony system: A Cooperative Learning Approach to the traveling salesman problem", *IEEE*

Transactions on Evolutionary Computation, vol. 1, no. 1, 1997, pp. 29-41.

[94] Carvalho P.M.S., Ferreira L.A.F.M., Lobo F.G., Barruncho L.M.F., "Distribution Network Expansion Planning Under Uncertainty: a Hedging Algorithm in an Evolutionary Approach", *Proceedings of tire 20th International Conference on Power Industry Computer Applications* (PICA 97), Columbus (Ohio), May 11-16, 1997, pp. 10-15.

[95] Yeh E.-C, Venkata S.S, Sumic Z., "Improved Distribution System Planning Using Computational Evolution", *IEEE Transactions on Power Systems,* vol. 11, no. 2, May 1996, pp. 668-674.

[96] Domínguez-Navarro J.A., *"Modelos para la Planificación de Redes de Distribución de Energía Eléctrica Asistida por Computador Aplicando Técnicas matemáticas Fuzzy"*, Tesis Doctoral, Universidad de Zaragoza, España, Marzo 2000.

[97] Miranda V., "Using Fuzzy Reliability Indices in a Decision AID Environment for Establishing Interconnection and Switching Location Policies", *Proceedings CIRED'91*, Lièje, Belgium, 1991.

[98] Mishima Y., Nara K., Satoh T., Ito T., Kaneda H., "Method for Minimum-Loss Reconfiguration of Distribution System by Tabu Search*", Electrical Engineering in Japan,* vol. 152, no. 2, 2005, Translated from Denki Gakkai Ronbunshi, vol. 123-B, no. 10, Oct. 2003, pp. 1149–1155.

[99] Young J., Jae-Chul K., "Application of simulated annealing and tabu search for loss minimization in distribution systems", *International Journal of Electrical Power & Energy Systems,* vol. 26, no. 1, Jan. 2004, pp. 9-18.

[100] Li K.K., Chung T.S., Chen G.J., Tang G.Q., "A Tabu Search Approach to Distribution Network Reconfiguration for Loss Reduction*", Electric Power Components and Systems*, 32: 6, 571-585.

[101] Das D., "Reconfiguration of distribution system using fuzzy multi-objective approach", *International Journal of Electrical Power & Energy Systems,* vol. 28, no. 5, Jun. 2006, pp. 331338.

[102] Sarfi R.J., Salama M.M.A., Chikhani A.Y., "A survey of the state of the art in distribution system reconfiguration for system loss reduction", *Electric Power Systems Research,* vol. 31, no. 1, pp. 61–70, 1994.

[103] Merlin A., Back H., "Search for a minimal-loss operating spanning tree configuration in an urban power distribution system*", Proc. 5th Power Syst. Computation Conf.*, Cambridge, UK, pp. 1–18, 1975.

[104] Shirmohammadi D., Hong H.W., "Reconfiguration for electric distribution networks for resistive line loss reduction", *IEEE Transactions on Power Delivery*, vol. 4, Apr. 1989 pp. 1492-1498.

[105] Borozan V., Rajicic D, Ackovski R., "Improved method for loss minimization in distribution networks", *IEEE Transactions on Power Systems*, vol. 10, pp. 1420–1425, Aug. 1995.

[106] Civanlar S., Grainger J.J., Yin H., Lee S.S.H., "Distribution feeder reconfiguration for loss reduction", *IEEE Transactions on Power Delivery,* vol. 3, Jul. 1988 pp. 1217–1223, .

[107] Baran M.E., Wu F.F., "Network reconfiguration in distribution systems for loss reduction and load balancing", *IEEE Transactions on Power Delivery,* vol. 4, Apr. 1989, pp. 1401–1407.

[108] Kashem M.A., Ganapathy V., Jasmon G.B., "Network reconfiguration for enhancement of voltage stability in distribution networks", *IEE Proc. Gener. Transm. Distrib*., vol. 147, no. 3, May. 2000, pp. 171–175.

[109] Chen C.S., Cho M.Y., "Energy loss reduction by critical switches", *IEEE Transactions on Power Delivery*, vol. 8, pp. 1246–1253, Jul. 1993.

[110] Nara K., Satoh T., Kitagawa M., "Distribution system loss minimum reconfiguration by genetic algorithm", *Proc 3rd Symp. Expert Syst. Applicat. to Power Syst. (ESAPS),* Tokyo and Kobe, Japan, 1991, pp. 724–730.

[111] Lin W.M., Cheng F.S., and Tsay M.T., "Distribution feeder reconfiguration with refined genetic algorithm", *IEE Proceedings Generation, Transmission and Distribution,* vol. 147, no. 6, Nov. 2000, pp. 349–354.

[112] Huang Y.C., "Enhanced genetic algorithm-based fuzzy multi-objective approach to distribution network reconfiguration", *IEE Proceedings Generation, Transmission and Distribution*, vol. 149, no. 5, Sept. 2002, pp. 615-620.

[113] Lin W.M., Chin H.C., "A new approach for distribution feeder reconfiguration for loss reduction and service restoration", *IEEE Transactions on Power Delivery*, vol. 13, Jul. 1998, pp. 870–875.

[114] Lin W.M., Chin H.C., and Yu G.J., "An effective algorithm for distribution feeder loss reduction", *Proc. IEEE Int. Conf. Power Systems Technology (POWERCON'98)*, Beijing, China, 1998, vol. 1, pp. 228–232.

[115] Huang K.Y., Chin H.C., "Distribution feeder energy conservation by using heuristics fuzzy approach", *International Journal of Electrical Power & Energy Systems*, vol. 24, no. 6, Aug. 2002, pp. 439–445.

[116] Liu C.C., Lee S.J., Venkata S.S., "An expert system operational aid for restoration and loss reduction of distribution systems", *IEEE Transactions on Power Systems, vol.* 3, May 1988, pp. 619–626.

[117] Jung K.H., Kim H., and Ko Y., "Network reconfiguration algorithm for automated distribution system based on artificial intelligence approach", *IEEE Transactions on Power Delivery*, vol. 8, Oct. 1993, pp. 1933–1941.

[118] Auguliaro A. et al., "Minimum losses reconfiguration of MV distribution networks through local control of tie-switches", *IEEE Transactions on Power Delivery,* vol. 18, , Jul. 2003, pp. 762–771.

[119] Hsiao Y.T., "Multiobjective evolution programming method for feeder reconfiguration", *IEEE Transactions on Power Systems*, vol. 19, Feb. 2004, pp. 594–599.

[120] Matos M.A., Melo P., "Multiobjective reconfiguration for loss reduction and service restoration using simulated annealing", *Proc. IEEE Power Tech´99*, Budapest, Hungary, Aug. 1999, pp. 213.

[121] Matos M.A., Melo P., "Loss minimization in distribution networks with multiple load scenarios", *Proc. IEEE Power Tech'2001*, Porto, Portugal, Sept. 2001, vol. 3.

[122] Matos M.A. et al., "Meta-heuristics applied to power systems," in *Meta-heuristics: Computer Decision-Making,* Kluwer Academic Publishers B. V., 2003, pp. 449-464.

[123] Pereira J., Saraiva J.T., Ponce de Leão M.T., "Identification of operation strategies of distribution networks using a simulated annealing approach", *Proc. IEEE Power Tech'99*, Budapest, Hungary, Aug. 1999.

[124] Khodr H.M., Ostos D., Mora B., Urdaneta A.J., Yusta J.M., Paiva P., Hussein H., "Optimal reconfiguration in large electrical power distribution networks", *Proceedings of the IV Venezuelan Congress of electrical engineering,* Caracas, Venezuela, Set. 2004, pp. 6.

[125] Ghosh S., Das D., "Method for load-flow solution of radial distribution networks", *Proceedings IEE Generation, Transmission and Distribution,* vol. 146, Nov. 1999, pp. 641-648.

[126] Gomes F.V. et al., "A new heuristic reconfiguration algorithm for large distribution systems", *IEEE Transactions on Power Systems*, vol. 20, no. 3, Aug. 2006, pp. 1373-1378.

[127] Gomes F.V. et al., "A new distribution system reconfiguration approach using optimum power flow and sensitivity analysis for loss reduction", *IEEE Transactions on Power Systems*, vol. 21, no. 4, Nov. 2006, pp. 1616-1623.

[128] Khodr H.M., Martinez-Crespo, J., Matos, M.A., Pereira, J., "Distribution Systems Reconfiguration Based on OPF Using Benders Decomposition", *IEEE Transactions on Power Systems*, vol. 24, issue 4, Oct. 2009, pp. 2166-2176.
[129] Morton A.B., Mareels I.M.Y., "An efficient brute-force solution to the network reconfiguration problem", *IEEE Transactions on Power Delivery,* vol.15, no.3, Jul. 2000, pp. 996-1000.
[130] McDermott T.E. et al., "A heuristic nonlinear constructive method for distribution system reconfiguration", *IEEE Transactions on Power Systems,* vol. 14, no. 2, May 1999, pp. 478-483.
[131] Goswami S.K., Basu S.K., "A new algorithm for the reconfiguration of distribution feeders for loss minimization", *IEEE Transactions on Power Delivery,* vol. 7, Jul. 1992, pp. 1484-1491.
[132] Kalantar M., Dashti R., Dashti R., "Combination of network reconfiguration and capacitor placement for loss reduction in distribution system with based genetic algorithm", *Proc. of the 41st International Universities Power Engineering Conf. (UPEC '06),* Newcastle upon Tyne, UK, vol. 1, 2006, pp. 308-312.
[133] Radha B., Ah King R.T.F., Rughooputh H.C.S., "Optimal network reconfiguration of electrical distribution systems", *IEEE Proc. International Conf. on Industrial Technology*, vol. 1, 2003, pp. 66-71.
[134] Chiang H., Jean-Jumeau R., "Optimal network reconfiguration in distribution system: part 2: solution algorithms and numerical results", *IEEE Transactions on Power Delivery*, vol. 5, 1990, pp. 1568-1574.
[135] Khoa T.Q.D., Phan B.T.T., "Ant colony search-based loss minimum for reconfiguration of distribution systems", *IEEE Power India Conference,* New Delhi, 2006, pp. 6.
[136] Mori H., Ogita, Y., "A parallel tabu search based method for reconfigurations of distribution systems", *IEEE Summer Power Engineering Society Meeting*, Seattle, WA, USA, vol. 1, 2000, pp. 73-78.
[137] Nara K., Shiose A., Kitagawa M., Ishihara T., "Implementation of genetic algorithm for distribution systems loss minimum re-configuration", *IEEE Transactions on Power Systems,* vol. 7, no. 3, (1992). pp. 1044-1051.
[138] Salazar H., Gallego R., Romero R., "Artificial neural networks and clustering techniques applied in the reconfiguration of distribution systems", *IEEE Transactions on Power Delivery*, vol. 21, no. 3, 2006, pp. 1735-1742.

[139] Das D., "A fuzzy multiobjective approach for network reconfiguration of distribution systems", *IEEE Transactions on Power Delivery*, vol. 21, no. 1, 2006, pp. 202-209.

[140] Kagan N., Oliveira C.C.B., *"Reconfiguração de redes de distribuição de energia elétrica através de ferramenta para solução de problemas de decisão com múltiplos objetivos e incertezas",* SBA Controle e Automação, vol. 9, no. 1, 1998, pp. 18-30.

[141] Shin J.R., Kim B.S., Park J.B., Lee K.Y., "A new optimal routing algorithm for loss minimization and voltage stability improvement in radial power system", *IEEE Transactions on Power Systems,* vol. 22, no. 2, 2007. pp. 648-657.

[142] Mantovani J.R.S., Casari F., Romero R.A., *"Reconfiguração de sistemas de distribuição radiais utilizando o critério de queda de tensão",* SBA Controle e Automação, vol. 11, no. 3, 2000, pp. 150-159.

[143] Taleski R., Rajicic D., "Distribution network reconfiguration for energy loss reduction", *IEEE Transactions on Power Systems*, vol. 12, no. 1, 1997, pp. 398-406.

[144] Dumbrava V., Comanescu G., Coculescu S., "Reconfiguration of the operation diagrams of urban electricity distribution networks by minimizing the energy losses", *16th International Conf. and Exhibition on Electricity Distribution,* Part I: Contributions, CIRED, (IEE Conf. Publ. No. 482), Amsterdam, vol. 5, 2001, pp. 5.

[145] Venkatesh, B. and Ranjan, R. (2006). Fuzzy EP algorithm and dynamic data structure for optimal capacitor allocation in radial distribution systems. *IEE Proceedings Generation, Transmission and Distribution,* vol. CLIII, no. 1, pp. 80-88.

[146] Silva Junior I.C., Carneiro Junior, S., Oliveira E.J., Costa J.S., Pereira, J.L.R., Garcia P.A.N., "A heuristic constructive algorithm for capacitor placement on distribution systems", *IEEE Transactions on Power Systems,* vol. 23, no. 4, 2008, pp. 1619-1626.

[147] Karmakar N.K., "Computational Results of an Interior Point Algorithm for Large Scale Linear Programming", *Mathematical Programming*, 52, 1991, pp. 555-586.

[148] Granville S., Melo J.C O., Melo A.C.G., "Application of interior point methods to power flow unsolvability", *IEEE Transactions on Power Systems,* vol. 11, no. 2, 1996, pp. 1096-1103.

INDEX

D

E

F

G

H

I

J

K

L

M

N

O

P

Q

R

S

T